WAR CAN'T STOP LOVE

To Erich

Enjoy the read!

Newzad Brifkani

WAR CAN'T STOP LOVE

The Story of Two Lovers Through Conflict.

NEWZAD BRIFKI

To arrange an interview or possible derivative works, please contact the author directly at www.newzadbrifki.com.

ISBN 978-1-7375843-0-8

To my brother and all the freedom fighters.

Contents

Acknowledgments

The first person I would like to thank is my brother, Ismet. This book is specifically dedicated to him and is based on some of his similar life experiences during the conflict between the Kurdistan Region and the Iraqi Central Government. I had the chance to interview my brother on several occasions. It was not only a chance for us to have a deep conversation but a chance to hear about the experiences he went through as a young man. Ismet was like a father figure to me, because we lost our father in the Kurdish-Iraqi conflict. Again, thank you Ismet. This book is for you.

Second, I would like to thank my mother for giving me information on how it was in her Kurdish village in the past, living through a war. As a woman, a widow, and a farmer, she gave me many details that I was too young to remember. Thank you, mother.

Lastly, I would like to thank all the other people that I had a chance to interview. I had the privilege to interview a few women from the Kurdish-American community who supported the Kurdistan independence struggle at the time. I also was able to interview some former Peshmerga besides my brother Ismet. These other Peshmerga, whom I will not name personally, gave me specific details of events that took place during the Kurdish-Iraqi conflict. They gave me details about their struggle to give their people more rights; their struggle to have an independent Kurdistan; the battles they went through; the weapons and tactics they used to defeat and eliminating the Ba'ath Regime. They gave me details about the days that they went hungry; months and sometimes even years that went by when they did not see their homes, wives, children, family, and friends.

God bless all of you for what you did to help the Kurdish cause, a struggle that is still ongoing today. Many of my interviews are voices of the martyrs. The people I interviewed told their Peshmerga experiences as well as civilian martyrs' stories. All real names have been omitted and will not be shared publicly. The story in this book is based on true events and timelines. To protect the identities of the interviewees, names of the characters are all fictional.

Thank you again and "Bejit Kurd u Kurdistan!" (Long live the Kurds and Kurdistan!)

Foreword

My first book, *My Journey to America*, inspired me to write this book, *War Can't Stop Love: A Story of Two Lovers Through Conflict*. When I was born in the Kurdistan Region of Iraq, there was conflict between the Kurds and the Iraqi Central Government run by Saddam Hussein and his Ba'ath Party regime. He was known to many as a dictator, since he was not elected and forcibly took the seat of president of Iraq.

The Kurds' story goes back to WWI, after the collapse of the Ottoman Empire. When the United Kingdom, France, and Russia divided the Middle East, the Kurds became a stateless people. Kurdistan was divided into parts of Southern Turkey, Northern Iraq, Western Iran, and Eastern Syria. Ever since they were divided up between those four countries, there has been conflict between the Kurds and the governments of those countries, which forbade them even basic rights, such as being citizens of those countries. In the Kurdistan Region of Iraq, a party known as the Kurdistan Democratic Party emerged under the leadership of Mustafa Barzani. He led a force known as the Peshmerga, which is translated as "those who face death." They are brave warriors that kept the Iraqi army away from Kurdish villages, towns, and cities.

As I was too young to remember much that had happened during the 1980s, my mother and older brother told us stories. This intrigued me to learn more. Who was I? Why was I in America? I knew we were living in a new country and a new place, but I had not known all the reasons behind this when I first arrived in the United States.

I knew the Kurdish people and the Peshmerga helped the coalition forces during the first Gulf War. Many years later, I saw the brave Peshmerga once again help the United States take out Saddam

Hussein, the dictator of Iraq, during Operation Iraqi Freedom in 2003. The Peshmerga emerged in the spotlight once again. This time, they came to the rescue of so many innocent people who had been oppressed by the rise of Islamic State in 2014. The Kurds were the boots on the ground, fighting to not only protect their homeland but the world by eliminating terrorism.

I sat down with my brother one day and brought up the idea of writing my second book. The second book would be a story of his life in the Kurdistan Region of Iraq, told as a love story, of two individuals during the conflict between the Peshmerga and Iraqi army. We sat down together many times to relive that epoch of his life and he tried to give me as much detail as possible about what he had seen and what we had been through. I used those interviews, along with the stories of other Kurdish community members, and combined them into what this book has come to be.

CHAPTER 1

My Love

She whispers in my ear, her warm breath tickling.

The hair on my body rises with jitteriness, my stomach is full of butterflies. "Stop it," I murmur, hoping she doesn't.

She whispers back, "I don't want this moment to end."

The sound of men's voices shatters that moment; farmers on their way back to the fields nearby. We jump to our feet and start running in opposite directions.

My name is Shivan. I am in love with a girl named Berivan. I am fourteen years old, and the eldest of my siblings. I have three sisters: Kurdistan is the oldest of them, then Sara, and Dunya. My two brothers are

Shervan, who is younger than Sara, and the baby of the house, Heja, whom we all love so much because he resembles our father.

I live in Belizan, a village forty miles from Duhok. Duhok is in the Kurdistan Region of Iraq. My family are farmers. We make nearly all our living by our farm, not just two or three kinds of crops. We don't get much stuff from the city because the Iraqi government makes it difficult to bring anything to the villages from the city.

We are Kurds from the north. Our language and culture are different from the people of South Iraq, where they speak Arabic. Most of the people of the north and south of Iraq are of the same faith, but we are totally separate races and ethnicities. We of the Kurdish north have been in conflict with the Arabic south ever since Kurdistan was divided amongst four very different countries.

Those countries—Iran, Turkey, Syria, and Iraq—have been oppressing my people, banning our religions, culture, language, and identity, trying to keep us from calling ourselves Kurds from Kurdistan. My father, Salih, owns and runs the only store in our village. It has a limited inventory, but we make a living off that, too.

Berivan and I have known each other since we were children. Her family is from another village, but came to live at Belizan because of the conflict and her father being a Peshmerga. She is the second oldest of eight siblings. She has three brothers and four sisters. Her family didn't have much in our village because they didn't own any land, though my father's father had given them an acre to farm. I didn't have any special feelings for her until recently. Here's how that happened...

I was wild. I didn't listen to anyone and my dad hated me for it. I was hunting birds, rabbits, and squirrels with my BB rifle one day when I got tripped by a rock and hurt my foot. It hurt so much I couldn't walk. I knew it was something serious.

Berivan was coming my way and saw me. She came running, asking if I needed any help.

Of course, I did. I had a big ego though, and wanted to act tough. "I am fine. I just need to lay down for a little bit," I said, sitting up straighter and brushing the dust off my shirtsleeve.

She saw sweat coming down my face, and I think I was turning yellow. There was a stream nearby. She grabbed her scarf from her head and went to the stream. She soaked her scarf in the cold water and came back, handing it to me to put on my forehead.

I held it against my head, thanking her as the cool wet cloth eased my clenching tension on that hot summer day.

Berivan stood, saying "I am going to the village to get some help." She looked so beautiful, especially after taking her scarf off. Her long brown hair flowed across her shoulders, her amber eyes like the shining sun, and a smile that showed her dimples. I just fell in love.

The next thing I remembered was waking up while I was being carried by my dad and uncles. I had a bad fracture in my foot and couldn't go out for a while. It was a great excuse because I could skip school. I hated school, especially because I had to travel to the city for my studies. It was in the Arabic language, which was foreign to me. It made me mad that I was not learning in my native tongue.

Saddam Hussein was murdering my people and we were forced to learn in his language. This made me sick to my stomach. Well, I guess we had no choice because Iraq was under his rule and he had made his native tongue, Arabic, the official language.

As days went by, I got bored inside the house. I decided to hang out in front of the house. This was a good choice. People of the village were constantly walking by. I said hello to them when they passed near our home. Occasionally I would hop across the road to my neighbor's house, pick a few of his apples, and hop carefully back. I would grab my sling shot and shoot at birds from time to time as well. Mostly I sat in front of my house and and enjoyed the scenery.

My village was beautiful. It was surrounded by hills, and Metin Mountain rose majestic just a kilometer north of us. On hot summer days we had plenty of fresh stream water to drink and to cool ourselves off. It was full of orchard trees and green grass. Our village had all four seasons, which was perfect.

Day by day, my foot was getting better. One day, as I was about to hop outside,

I saw her walking by. Berivan was right there! I tried to hurry but was too slow to catch up to her. *Dang. I missed my chance to say hello.*

You see, in my culture, it's not normal for a guy to talk to a girl alone. In a small village, everyone knows each other. They could think negatively about such chatting, and it's not part of the traditional way of life for a boy and a girl, or even a man and a woman, to have private conversations. It's better if there are other people around. But having other people around makes it difficult to tell a girl that you are in love with her.

The next day I went out and explored a little, testing the strength of my bad foot, but really looking for Berivan: nothing. I did not see Berivan anywhere.

The day after that, in the mid-afternoon. I saw Berivan coming up the road that passed by my house. I was standing in the shade by our front door, leaning my back against my house, propping my bad foot up a little in front of me to give it a rest. This was my chance to speak to her, to see her golden eyes holding my gaze again.

However, there were a couple of other

villagers behind her. I was shy to say hi to her in front of the others, but also determined to do it. As casually as I could manage, I stood up and limped to the road in front of her before she passed by. But for some reason, I looked straight over her shoulder and greeted the other villagers, who were still at a distance. Berivan seemed surprised, but since I could not bring myself to look at her face and lose myself in those golden eyes, I couldn't be sure. She walked right by as I stared at the other villagers, as if waiting for them to reply.

As they got closer, they asked if I needed any help. I was now doubly embarrassed and said no. I told myself, man, that was way too soon to say hello.

Later that evening, I started to think about a better way to approach Berivan. How would I manage it? What would I say? By the time I fell asleep, I had a plan. Luckily, I had small siblings. I just needed to convince them to help me.

The next day, I asked my younger sisters Sara and Dunya and my younger brothers Shervan and Heja if they wanted to go play outside. They said yes, and it worked: we had fun and attracted attention... but there was no sign of my love Berivan. My siblings finally got tired and wanted to go inside.

I was undeterred. I told my sister Sara, to wait in the patio and play there, where she could see the road. If she saw Berivan, I told her to let me know before Berivan got close to the house.

She agreed to do it if I would give her a gift.

I agreed.

About an hour later, Sara called my name. I went to check on her and she pointed to the road: Berivan was approaching. I went outside as quick as I could. Luckily, no one was walking with her or behind her. This was my chance.

She drew near our house and looked sidelong at me. She had a smile on her face, and I realized she must have heard Sara calling to me. As she was walking past, she said playfully, “You are not healed yet?”

I returned her smile. “Now that I have seen you, I’ll be better.”

She turned her eyes to focus straight ahead and kept on walking, but her smile was wider than before.

I had finally connected with her.

Days went by and I saw Berivan more often. We started to talk a little more each time she walked by my house. It was not a conversation, just small talk: Hi, how is your day? You look nice today. Berivan was educated, too. She was going to school in the city of Duhok. I was going to the city of Zakho, fifty minutes from Duhok.

We started to write to each other in Kurdish. This was a better alternative for conversation because we could not speak together publicly. Writing gave us a chance to say more of what was on our mind, and share the feelings that we had for each other.

In her letters, she wished and prayed for more. She wished for peace and that the war would end. She wished to live like the people in the city: having a nice house and a life like many normal people did. Berivan was very patriotic. She praised the Kurdish fighters, the Peshmerga. In her letters, she told me that she would fight too if they let her. She wanted to get revenge for some of her family members that had been killed by the Iraqi regime.

In my letters to Berivan I described

how my father was upset with me. He was constantly pressuring me to go back to school, or even to live in the city with my other relatives until my education was finished.

I told Berivan how I felt about her; that one day, I would ask for her hand in marriage. When I mentioned this, her letters came back to me insisting that I listen to my father and do what he said. People looked up to him as being a wise man, and not just in Belizan village, but in most of the Duhok area. Berivan also told me that she loved my father for what he had given to her family. She was happy that I wanted to marry her in the future. She replied, "Taking my hand is not going to be easy. My parents hate your other relatives, but they do like your parents."

In the Kurdish culture, when marriage happens, you don't just get married to the bride but to the whole family. When a man chooses to get married they had to be serious. Getting divorced after marriage was taboo. If I was serious, I could not back out. That is how Berivan and I started to fall for each other.

We both had an education from the

city despite being farmers, even though it was in the Arabic language. When we were home, we had local tutors who visited our village from time to time. They taught us Kurdish, as it was banned in the city's education system.

After about a month, my foot was still only slightly better. I could only take small steps without pain. My father came to me one day and said, "I have a Christian friend in the village of Baze, we will go see him." Baze village was a twenty-minute walk from Belizan. His Christian friend was named Oshane, and known by many as the medicine man. He healed bones that were fractured. He was very famous amongst the Kurdish villages in the Duhok area.

While my father chatted with Oshane, he asked how I fractured my foot. I explained, and he pretended to scoff at my choice of prey. He said, "birds barely have any meat on them. Why don't you hunt boar, they have much more meat!" He laughed uproariously at his own joke. He knew that we didn't eat pork.

He took my foot in his hands, massaged it for a while, fetched some herbs that had been soaking in water, and rubbed them on my foot. He then produced a plain-looking stick, about the length and thickness of my largest finger. He said, “Hold this stick between your teeth.”

I put the little stick in my mouth and turned to my father —and Oshane suddenly twisted my foot hard. It felt like he pulled my foot out of my body. “You are done,” he said.

It felt like he had fractured my foot again.

As we were walking back home, I limped most of the way. But before we reached our village, as the sun was setting, I started to walk normally. The pain was mostly gone. The strength was returning. I was amazed. I turned to my father and asked, “Is Oshane related to Jesus?”

My father smiled at me and said, “Oshane is a good man, he is a loving person and a Kurdistani patriot.”

Perhaps this meant Jesus, too, would be a Kurdistani patriot, since Jesus was a good and loving person like Oshane. And both of them worked miracles. I smiled up at my father.

My father patted me on the shoulder. "Now you can go back to school."

The next morning for some reason I woke up extra early, before the imam in the village called to prayer. It was still dark out, but I could hear the dogs barking, the crickets chirping, the wind so soft, fluttering the leaves of the trees. I could see the sun, barely making its way up behind the shoulders of the mountains.

Our village was surrounded by mountains, especially big Metin Mountain that stood high and sharp against the sky, southeast of our village. Kurdistan is not like southern Iraq. We have mountains, trees, green grass, lots of fresh water, and all four seasons. In winter it gets cold, and we get a lot of snow. Spring is the most beautiful season. There is lots of rain, and flowers of every kind you can image bloom everywhere. In the summer it gets really hot but the heat is dry. All one needs is shade and it feels okay. During fall season the weather is cool and all the orchard trees are ripe and ready to be picked.

This morning the sun seemed slow to rise, but I waited for it. It came up at last, and the moment was etched into my memory forever. I felt the morning sun's warmth caress my face and I smiled. I thanked God for what I had, and for what I wanted.

Before my family woke, I started a fire outside, got the tea kettle ready to make tea, and went to the stream by my house where my father had made a concrete water station. He had worked on it for weeks and made it accessible for everyone in the village to get water, coming from three wide steel pipes that jutted from its side.

I washed my face and the cold water woke me up. I put my kettle by the nearest pipe to fill it up with water. I saw a bucket going next to the metal pipe by mine. I turned and looked up. It was Berivan.

Just like the sunrise I had just witnessed, just like the twenty rays of the Kurdish flag, she smiled and said, "Spede bash" (good morning). I looked at Berivan for a second, unable to speak, then snapping out of my stupor I smiled and said, "Spede bash!"

We talked for a while after our kettle and bucket had filled. I could hear the neighbors nearby, just on the other side of the water station, getting ready for prayer at the mosque. I asked Berivan if she wanted to go somewhere alone with me.

She said no at first. She had to get breakfast ready for her family. Her father was going to the mosque to pray.

I looked down and nodded. "Okay." I shifted the kettle to my other hand, rose to my feet and turned to go.

As she walked with me along the path to the road, she asked, "How long do you want to see me?"

"As long as it is safe," I told her.

She said, "Meet me at Kani Hishke in one hour."

Kani Hishke meant Dry Well. The well wasn't really dry. We called it the Dry Well because sometimes the water level got extremely low in it because it was so deep below ground. Most years, during the summer months, the water was low in the well. This was why the villagers called it Kani Hishke.

I could not believe she had agreed to meet me somewhere alone. As I was walking home, I was so excited, I dropped my kettle full of water. That did not bother me, I didn't need tea. I threw the empty kettle by the house and started to run to Kani Hishke.

By the time I arrived at the well, the hot morning sun was high above the mountain, burning my face. It felt good, yet I was sweating because I was so nervous. The sun just made it worse. I leaned over the well to scoop up some water. The well was full so it was within reach.

I grabbed some water with the palm of my hand and splashed it over my face. It felt cool, refreshing. I scooped another handful and brought it to my mouth. The color was so clear, and the taste was so good.

Someone grabbed me from behind and nearly pushed me into the well!

I regained my balance and turned around to find Berivan, laughing!

"I could have thrown you in there just now."

"Yeah, right." I made my voice sound scornful but I was smiling so widely I didn't fool her. I took her hand and we went to sit in the shade of the tree by the well. It was a walnut tree, old and broad.

It was too early for us to pick the walnuts, but the leaves gave plenty of shade.

Berivan and I talked about when we were kids and the jokes we used to play on each other. We laughed and tickled each other as the conversation went on. I turned around and faced her. I leaped over her body. I felt good as I got her close to me. I felt secure, in peace and comfortable. Like I was holding a pillow, except it was for real. I kissed her on the cheek and told her I liked her a lot.

She asked if I had feelings for anyone else. I said no, right away. I mean, in my mind I thought there were a few pretty girls in the village. I can't lie about that. I could not tell Berivan that though, and said no right away. I was being honest when I said no. Even though I thought there were other pretty girls in the village, and some of them I liked very much, my feelings for Berivan were different.

I liked Berivan a lot, so much that I wanted her to be my future wife. The other girls I had just wanted to fool around with, if provided the chance. There is a difference between attraction, affection, and love. I know she didn't believe me, but I knew that I wanted to marry her.

That was what was important to both of us.

I brought her closer to my body. The warm wind caressed our skin and set her hair to waving and curling around her face. Her smile was to die for. She looked so perfect, as if God had drawn her with a pen. Softly, I kissed her neck. She tried to back away, but I held her tight. She relaxed in my arms.

I slipped my hand under her dress. She did not stop me. Berivan seemed to like my company. I felt like I was in heaven. We fooled around for a while, talking to each other, whispering in each other's ears. The day seemed perfect and thank God there was not another villager around. It was still early though, and work was not started yet.

I could hear the barking of the village dogs, the roosters crowing. I saw smoke going up in the village. A sign that the villagers were waking up. I did not care though. I did not want to let go of Berivan.

I grabbed her again and kissed her soft cherry lips. She closed her eyes when I kissed her lips. I told her not to close her eyes. I wanted to stare at them because they were so beautiful. She said, "I don't want this moment to end."

Then a sound of men's voices came carried on the wind. Villagers were coming our way.

Berivan got up quickly. "What's that, where are they coming from?"

I knew where the men were going. It was time to work on their crops. They were heading to the fields. It was not our way to be found like this. The best thing for us was to leave as quickly as we could.

We hugged and said goodbye. Then we ran in opposite directions. I ran towards the men headed out the west side of the village, towards the fields, and she went north towards the village itself.

CHAPTER 2

Peshmerga

My father and I were in the field one summer day, planting our crops, along with my Uncles Azad, Rojhat, and Checkdar. Azad was younger than my father by a few years. He was about 5'8" in height, medium built with light tan skin and black hair. Rojhat and Checkdar were a few years younger than Azad. They were twins. Very handsome looking, they had light tan skin, dark brown hair, medium build but both a couple of inches shorter than Azad. But although they looked very similar, Rojhat and Checkdar's personalities were very different. A person could tell them apart by the expressions on their faces alone.

The sun shone hot upon us with no cloud in sight near Metin Mountain. The usual sounds accompanied our work: birds chirping, dogs barking, livestock sounds from the cows, chickens, roosters, donkeys and sheep. A calm wind rustling in the trees and giving us some respite from the heat, until I heard children shouting. It came from near the streams. They were calling out, "They're here, they're here!"

Who was here? I asked myself. Then I looked upon Metin Mountain and my eye caught them. Many trees concealed them as they descended the lower slope of the mountain, but I counted about thirty men. They looked sharp. Warrior style, with baggie pants, military vests. Each had a rifle on their shoulder. They wore Kurdish turbans, some black and white and most red and white, representing the Kurdistan Democratic Party (KDP) Peshmerga. They were headed towards our village.

I was used to seeing men wear similar Kurdish clothing, but these men stood out. They looked brave, like lions coming down to eat us in the village. I said "Dad, look there."

He spotted the men immediately. Looking to my uncles, he called them by name to get their attention: "Azad, Rojhat, Chekdar, we have to get back to the village."

My uncles followed my father's gaze and saw the men descending the mountain. Lion-men. They put down their shovels and told my father, "bele tu rast dibeji" (yes, you are right).

My father was the oldest of the brothers. He had a sister once, but she died when she was young. As I said, my Uncle Azad was next in age after my father. Rojhat and Chekdar were twins and younger than Azad. Since my father was the oldest, they listened and looked up to him. Not to mention the brains my father had. The people of our village always liked my father, but not my uncles. The villagers would tell my father that my uncles were troublemakers, and there was some truth to this. My uncles were always picking fights with them. My father would laugh, but they didn't find it funny. They would complain to my father. My father would smile and say, "I will talk to them." The villager would walk away really quickly in anger.

When the Peshmerga would come, the villagers would host them. The Peshmerga moved around a lot. They had no permanent bases near any villages for the safety of the civilians. The Iraqi army was the fourth largest standing army in the world. They had all the modern weapons to fight the Peshmerga. To avoid jet bombings, the Peshmerga moved from village to village. Now, these Peshmerga were coming to ours.

We got to our village before the Peshmerga did. My father told my mother Aysha to make more supper than she was planning to. My mother was really beautiful. She had long black hair and light skin. She was younger than my father about ten years. My mother knew immediately what my father meant about making more supper than expected. You see, my father was in command of information to the Peshmerga. The Kurdistan Democratic Party had entrusted him with the rank of civilian commander. He was a Peshmerga, but undercover. No one was supposed to know. He passed along information to the Peshmerga coming into our village from others who had visited the village before them, and from them to those that would come after. He would give them the status of the fighting, the injured, the killed, and news from the Federal Iraqi Government.

The Peshmerga were very well trained. They had manners, honor, and courage. They did not want to intrude in our village. They would send out one scout. He would meet with the village mayor. That one Peshmerga would ask the mayor for permission to stay the night in the village.

Our mayor at the time was Suleyman. Suleyman was a good man, but his followers in the village were mostly bad. They did not like my family. As the Peshmerga scout walked by he asked for directions to the mayor's house. I told him it was just past the mosque, the last house in the corner.

Uncle Azad was a very impatient man. He overheard the Peshmerga asking me where the mayor's house was. He hollered at the Peshmerga, "Hey, hey, where are you going?"

The Peshmerga took his rifle from his shoulder and started walking back towards my uncle Azad. Gripping his rifle tightly, he stared my uncle in the eyes. "What did you say?"

Uncle Azad spoke boldly, gesturing toward my house. "Your commander is in here. I am also a Peshmerga."

The Peshmerga shyly lowered his rifle. He blushed with shame and humiliation. He jabbed his hand out and walked towards Azad to shake his hand.

Azad quickly grabbed his hand and shook it hard and fast.

The scout said, "I apologize, but these are my orders: to see the mayor first before I talk to anyone else. These are my orders," he said again.

Azad looked at him, smiled, and said, "Let's go, then."

Azad and I walked with him to the mayor's house. After visiting with the mayor, the Peshmerga scout told the mayor that he was with Saeed Chelki's brigade. Saeed Chelki was a short, heavy built man with a mustache. He was known to be a warrior and a very skilled fighter. Most of them knew each other, as they were from the same tribe. He told Suleyman that they had thirty men, including himself. They wanted to be divided amongst the villager's homes. They would appreciate it if the villagers hosted them for the night.

Our mayor Suleyman said, "Yes, of course. I will inform the other villagers and have the houses ready by the time you inform your unit that they are welcome here."

The Peshmerga left, leaving me and my uncle behind. My uncle soon excused himself from the mayor's house and we returned to our house again. He told my father what happened.

My father said, "Azad, let the Peshmerga do what they need to do. Leave people alone. If they need help, we will help them. We don't need to reveal ourselves or show off. If they are knowledgeable, they will end up at the right place."

We saw the mayor walking towards the mosque. He entered, got on the loudspeaker, and called all the heads of households to meet at the mosque.

Of course, all the heads of households were men. My father told Azad to go get his other brothers, Rojhat and Checkdar, and tell them to meet at our house prior to going to the mosque. We waited about half an hour for my uncles to return to our house. My father told them, "Whatever the mayor says, you must listen. Whether it's one Peshmerga or three, you must welcome them and host them the best way you can, with food and shelter."

My uncles agreed and said they would do what he said. At that moment, there was a knock on the door. My father went to open it and there he was: Saeed Chelki from Berwari Bala. He was the head of the Peshmerga unit visiting our village, and he had come to visit my father. Saeed was a very tall man, over six feet in height. Long thick black hair escaped from under his Kurdish turban. He shook my father's hand and gave him a hug. Laughing, he said, "I am sorry that my Peshmerga did not come to your house first. I thought the villagers would direct him to your house, not the mayor's."

My dad smiled and nodded. "Please, come in."

Saeed Chelki told his Peshmerga to go on to the mosque, except for him and three others who would stay here, at my father's home.

That night, I stayed up as long as I

could. I listened to all the conversations that my father, Saeed, and the other Peshmerga had. My uncles would join the conversation from time to time. However, my father was the one with the jokes and long stories. I could see in the eyes of Saeed and the other Peshmerga the respect they showed to my father.

It was a long conversation. I brought food to them, followed with tea, three or four times. When you bring tea three or four times that means that the guests are comfortable. My father pulled his tobacco from its sack. I remember the strong smell because I used to steal some from him and smoke it with my friends. He passed his tobacco sack around. Saeed rolled a cigarette as did one other Peshmerga. All my uncles did too, as they barely had any of their own tobacco. When they did have tobacco, they didn't have the paper to roll it up with. My father, living comfortably in our village, always provided my uncles with tobacco and rolling paper when they needed it. Not to mention food and other household items that they could not afford.

I went to the corner of the living room. There was a pillow there. It was very comfortable. There was smoke in the living room from the cigarettes, and warm light from the wood-fire burning in the chimney-stack in the middle of the living room. My stomach was full. I looked at Saeed. He produced his own pack of cigarettes to share with the men, and I blinked in surprise.

You see, it was tough to find cigarette packs. Most villagers didn't have the money to afford them. The Peshmerga were mostly volunteers, and had less money than any villagers. I assumed they must have grabbed them off an Iraqi base they had captured, and maybe they had gotten good rations this month. My father had told us that the Peshmerga rarely received supplies. Heck, they barely had any weapons to fight the Iraqi soldiers.

The soft conversations continued as my eyes blinked slowly, and slowly, my body at rest, I drifted into sleep.

Next thing I knew, I heard voices coming to my ear. I slowly opened my eyes and saw that my mother was making breakfast. I looked outside: it was still pitch black out. When the Peshmerga would come into our villages, they would often come before dawn and leave before dusk. They did not want to be spotted by Iraqi planes.

I heard my mother calling me. I got up and went to her. She said, "You have to help me get breakfast ready." I took a few items from her that she made and began to arrange them on the table. They were eggs, tahin (crushed sesame seeds), mast (Kurdish yogurt), olives, fish, potatoes, afka bajana (Kurdish tomatoes fried with seasonings), and of course fresh Kurdish bread.

Saeed Chelki's Peshmerga gazed in awe. They were in disbelief that my father would host them like this. Many of these items were rare, not to mention how poor most villagers were, and most couldn't afford to have them.

My father smiled. He was often surprised by my mother. My mother always told him, when he wanted a good meal, that they didn't have the supplies, or the things he was asking for weren't available. But when guests were over, there was always plenty to feed them. My father was very happy with my mother. Even though she was more than ten years younger than him, she always acted mature.

My mother had a tough upbringing. Her parents had died when she was young. Her mother died in childbirth; her father was killed by wrongful accusations. She grew up as an orphan, raising two of her younger sisters by herself. She was from another village in Berwari Bala called Qumriye, but her father had come to Belizan for work and stayed here ever since. When her father died, her aunt took care of her until she was old enough to get married.

My father told me that many people came to ask for her hand. She was a good-looking young woman and a hard worker. My father said this second quality was important for parents who were looking for brides for their sons.

My father and mother liked each other. Nothing like what Berivan and I have, but they knew there could be some kind of love there. My grandpa encouraged my father to ask for her hand. My father obeyed, as this was the right thing to do: please your parents.

My mother had two younger sisters and didn't want to leave them behind if she were to get married outside of her ancestral village. Who would take care of them? My mother told her aunt in Belizan that she did not want to marry anyone outside of her ancestral village, Qumriye, because she wouldn't be able to take care of her younger sisters if she did that. My mother agreed to marry my father under one condition: that she could bring her sisters to be with her in Belizan, to be taken care of until her sisters were old enough to get married too. My father agreed, and that is how they were married.

The Peshmerga finished their breakfast. My father called me over. I came to the living room area and picked up the dishes. The Peshmerga looked at me while Saeed Chelki patted me on the back with his cigarette in his mouth, saying, "Work hard so you can be like your father."

I looked at them, smiled, and said "Bele, mam" (yes, uncle).

The mosque speakers blared "Rabin, rabin" (wake up, wake up), which was a signal for the Peshmerga to meet at the mosque, to gather to leave the village before daylight. Usually the imam calls us to prayer every morning before any noise is ever heard from anyone else. Instead an announcement was made by the Peshmerga first, followed by the imam calling to prayer.

My father and many of the other elders in the village knew and understood the message. They each came out of their houses with the Peshmerga to the mosque. They started to say goodbye to each other, shaking hands, hugging, and kissing each other on the cheeks, a traditional Kurdish greeting or goodbye. Then, slowly, the Peshmerga disappeared into the woods and out to the fields. The Peshmerga headed for the mountains during the day. That is where they would go to seek shelter, attack Iraqi troops, and visit another village when night came.

I wanted to leave with them so bad. I wanted to grab my dad's rifle and join the brave Peshmerga. These were men that faced death. They were fighting to free the Kurds, in hopes of winning greater freedoms for us and someday an independent nation of our own, called Kurdistan.

I asked my father about it now. "Dad, I want to be a Peshmerga."

He looked at me. "Your education is more important."

"Well, what about our fight? Our fight for freedom, and to have our own country too?"

He glared at me and yelled, "You are just a baby. You think it's easy being a Peshmerga? They are volunteers, their salary is shit. They go days without food, they barely have enough ammunition to defend themselves or pick a fight. Now shut up and sleep early tonight. I am taking you to your uncles in Zakho, so you can go back to school."

"I hate school. I don't want to learn those camel fuckers' language. I want my country back. I want my freedom. I am joining the Peshmerga. I am not going to Zakho."

I ran home from the mosque.

I came home mad.

My mother asked, "What's wrong?"

I ignored her and laid in the living room with a pillow to my face.

My mother did not bother me with any more questions. That was what mothers did. They were always there, but never intruded in the males' lives. She assumed that if I needed something I could talk to my father.

That was not the case, though. I did not get along with my father.

He wanted so much for me. Especially an education, which I knew would get me nowhere. I do admit that I had become very good at the Arabic language, having gone to school all these years. I was able to communicate well. When I would come back to the village from Zakho, visiting Iraqi soldiers would talk to me and would be impressed with my Arabic skills. I hated them but didn't show it. I made them laugh from time to time. I was always talking about girls with them. They loved these conversations; they were horny fellows spending a long time away from their homes and wives.

I needed someone to understand. It was my time to fight. I wanted to be a Peshmerga, a freedom fighter, fighting for a just cause. If I were to die, I would be considered a shehid, a martyr. Our holy book, the Quran, says, "those who die fighting to defend their homes are shehid." If I died trying to save my people, I would go to heaven anyways, I told myself. I was defending my homeland, and this was easily going to be granted to martyrs who were killed defending their homeland. When you kill someone on your land for no reason, that is murder, but when you kill others on your own land because they are trying to kill you, that is self-defense.

Besides, I would be fighting amongst men of honor, bravery, and virtue; men that women talked about in the village constantly. Berivan would surely be happy and supportive of my cause if I joined the Peshmerga. I had thought about it for a long time and had my mind made up. I had looked at my situation from all sides, and choosing to become a Peshmerga would be a relevant choice I made for myself.

My father came home late that evening. I could hear him enter the house, but I was too tired to raise my head up. I could hear his gentle cough. He moved quietly in the darkness so as not to wake my siblings and mother. He did not turn on a light because we had neither electricity nor heat.

The Iraqi government did not treat the Kurds as part of Iraq. They only cared about our rich lands with lots of natural resources. Things like electricity or natural gas service were only available in big cities. Small towns and villages were left without them. This was also a tactic used to prevent the Peshmerga from living well.

Most of the Peshmerga hid in the mountains. They stayed away from the city. They knew it was too dangerous, that traitors would point them out to Iraqi authorities. Our lamps were not "switched on" but "lit" to give us light. I could hear my father get out a tin of kerosene. He filled the lamps and lit one of them, and I could hear him add some wood to the chimney-stack to give us heat. That is how we had light and heat.

I woke up early in the morning the next day. My father was awake before me. He was sitting in the front porch, gazing at the mountains, with his spiritual beads in his left hand and his cigarette in his right. He had a pot of tea on the fire pit away from him. With his tea in front of him, steam still rising as if he just poured it, he looked amazing.

My father was a good-looking man. He had long enough dark brown hair that he combed it from left to right. My father was about 5'8" in height, with beautiful light tan skin and a thick dark mustache. He wore bergiz, the expensive and stylish Kurdish clothing for men. Many people could not afford bergiz. He had a few pairs and most of them were dark blue in color. He had his Kurdish turban next to him. He always looked ready and sharp, like he was going into battle.

Yes, people respected him. They respected him because he was home-educated, mostly by Peshmerga who taught him at an early age how to read and write. Some of his Peshmerga friends also taught him how to read and translate the Quran. He had wise men as friends.

My father learned from their experiences and stories of the past. He loved peace. In our village, when fights broke out between the villagers, he would always step in to prevent further escalation. These were some of his good traits that people knew him by. People wanted to be like him. I wanted to be like him. I just wish he understood what I wanted, and accepted that.

I came to him and said, "Spede bash" (good morning).

He replied, "Bashter" (a better morning to you).

I smiled, took a glass, poured some sugar into it, and went to the fire pit. I lifted the tea pot and poured myself some. I picked a thin stick from the ground by the fire pit and used it to stir my tea. I came to sit by my father.

He laid his beads in his lap, retrieved his cup of tea, sipped thoughtfully, then looked at me. "Son, are you ready to go to Zakho? You know, school is important. One day you will need to use this education to fight for our country. We must not only learn how to use a gun, but also a pen. There will be a time when they will give us a chance to express ourselves through debate and not fighting."

I asked myself, *how does my father know all this stuff? What if they kill us all before they give us a chance to ask for our rights through a pen and a debate?* I was afraid to respond this way to my father, so I simply nodded. "I will be ready in a couple of days. I just wasn't feeling too good." It was a Wednesday.

Thursdays were half days and on Fridays there was no school.

He said, "Fine then. I will take you early Saturday morning. You can stay with your Uncle Omar and come home to visit every Thursday to Saturday morning."

I agreed. I could hear my other siblings waking up. Some of them began crying and now the others were screaming at each other. My mother came out and said, "I will get breakfast ready."

I grabbed a container and went to the water station to bring some water for our breakfast.

As I was walking, the Peshmerga longing hit me again. I asked myself, *how do I join the Peshmerga?* Ideas came to me: I could run off and find a unit to join. I could ask fellow villagers if they had some contacts or connections nearby, to give me an introduction and recommend me to a Peshmerga unit.

I shook my head and erased that idea right way. I knew I couldn't let anyone know of my plan. If anyone knew what I was up to, they would tell my father and it would be the end of that. The best thing was to run off and find a Peshmerga camp, I told myself. I would say *I am here ready to fight*. They would give me a rifle and I would be recruited.

The Peshmerga were not easy to find, though. They were always moving from one village to another. Their bunkers were temporary. When I would go with my father to the mountains to hunt, we would find a lot of them empty, and no one in their bunkers. They would leave some of their belongings there though, with their fire pits and cans of food laying around. Their bases and bunker areas were covered with big rocks to protect them from bullet fire.

I came back with the water containers filled. My family and I had breakfast. My father said he was going to the field to work on the crops and orchard trees.

This was perfect. I would leave to join the Peshmerga that afternoon. Before I would leave though, I would go to see Berivan. I had to say goodbye. I was too afraid to tell my mother, though. She would not agree with this one bit.

I told my sister Kurdistan to go find Berivan. She knew that I liked her. I told her, jokingly, that if anyone knew my feelings for Berivan that I would kill her. She was my closest sibling and very supportive. I was happy Kurdistan had agreed to go and let Berivan know I wanted to speak with her, to tell her to meet me at our store. Kurdistan agreed, for some jewelry. Of course, I would take the jewelry from my dad's store.

I told my mother I would open the store up until father came back from the fields. A few customers came to the store that morning and I made some sales. Then Berivan came. I knew she did not have any money, her family was poor. She was not here to shop. She was here to see me.

No other villagers were in the store, so she said, "hello, Shivan."

I grabbed her and hugged her.

She pulled away enough to look into my eyes. "What's wrong?"

"I am joining the Peshmerga."

Half-smiling, she said, "What? You are kidding, right? You can't, you are too young. Your father would kill you."

"I have made up my mind." I gave her a kiss on the cheek. Then I gave her a bracelet that I had taken from my mom's jewelry box. Using the diminutive form of her name, I said "Beri, I will come back, and we will get married."

She looked down her feet sadly. I grabbed some tea and put it in a bag with some sugar. I also put some tahini, olives, and cigarettes for her father, and said "Please take this for your family." It was nearing the afternoon. I had to leave the store. If I went back home for lunch, I would be stuck.

I started along the road towards my house. Instead of stopping there, I walked right by it and kept on walking toward the Metin Mountain, which overshadowed our village from the south.

I had about five hours before sunset, before it would get dark. I started to pick up my pace. I had one intention, and that was to join the Peshmerga. I didn't know how I was going to do it. I was just going to ask them when I approached them. Goosebumps prickled all over my body. As I was walking, I started to ask myself some questions: *Which way do I go? Do I go West towards Babire village?* I would assume that Peshmerga would be there. There was the river that ran to the city of Zakho. Peshmerga needed water, especially a river. Also, Saeed Chelki's village was nearby. *Do I go east towards*

Baze Village? This is where my dad's friend was, the Christian medicine man, Oshane, who fixed dislocated and fractured bones. *Do I go south towards the city of Duhok?* Along that way there were many villages and I would be sure to run into Peshmerga. There was only one catch: I would have to cross the Metin Mountain, and that was not going to be easy.

I didn't have much time. It was a mountain high in elevation and many wild animals were there, including wolves and bears. I had to find the Peshmerga before nightfall. I did not have fire, food, or a weapon.

Going up the Metin Mountain unprepared was a dangerous move. The good thing about Metin Mountain was that it had many caves. Lots of Peshmerga had bases there. I would be sure to find them. If not, I would hide in the caves at night. When morning came, I would continue to look for them again.

I took my chances and went straight south. This would be a great risk, but it seemed to be the best way. I was sure that I would find the Peshmerga bases and join their units.

As I was walking, I saw local villagers. They asked where I was going. I told them I was just hunting for birds. I did not have my sling shot with me. I only wore khak pants and a long-sleeve shirt, the traditional Kurdish clothing. I knew that they saw red flags, but they didn't bother pestering me since it was going to be dark soon so they didn't think I would be going far.

On my way, I picked apples that were barely ripe and other fruit from the orchard trees that I came across. There is a Kurdish saying, "if you see an orchard tree, pick as much fruit to eat as you want for the moment, just don't take extra, which would be a sin." I knew I needed food for my journey, not just for the moment, so I committed a sin and filled my pockets with apples and yellow cherries.

I continued to walk and was about two hours away from home. I turned around and saw my village spread out far below, at the feet of the mountain. I was halfway up already, but I could see the sun would be setting soon. As I walked, I looked around, but could find no Peshmerga. I saw a couple of their bunkers, but they were empty. They had fire pits next to them, but no wood. They had empty cans laying around and other small empty wrappers of food. I asked myself, *should I stay here?* But I knew it was a bad idea. Even if I could find enough wood to stay warm overnight, I had no lighter or matches to start the fire.

I continued to climb until I could see that the sun had finally set, and it was getting to be dark. I rested by a great rock outcropping and just gazed around. I had to find some sort of life that could help me. Suddenly, I heard sheep in the distance.

I looked around and saw that there were sheep grazing along a less-steep area of the mountain slope to the east of me. Perfect! Where there are sheep there are dogs, and where there are dogs there is a shepherd.

I started to run toward the sheep as fast as I could.

As I got closer, I could hear the dogs begin to bark. I knew they could smell me, but I was not a wolf and they would not kill me. The worst the dogs would do to me is knock me down and bite my arms and legs, perhaps my face and hands if I were foolish enough to let them. From their bark I could tell the dogs were Kurdish Kangals. They were big and fierce, and often kept by shepherds as working dogs, as protection from wolves and bears as well as helping to herd the sheep. If they were not trained right, they would be very mean and aggressive.

I started to holler to the shepherd, "Shivan, Shivan!" It felt like I was calling my name, but it really meant "shepherd" in Kurdish.

After I had called out a few times he responded "Bele!" (yes) Following the sound of his voice, I saw him rounding another rocky outcropping like the one I had just left. I slowed to a walk and turned towards him as he did the same. The dogs started running at me, barking, but at a shouted command from the shepherd they ran the opposite direction, back to the sheep.

I finally reached him. I did not recognize him. He was not from my village. He was a short young man who appeared to be in his late thirties. His clothes looked like he had been attacked by wild animals and ripped up by tree branches.

He looked me up and down with an

expression of mild surprise and said, "Are you lost?"

"No. I am trying to find a Peshmerga base. Where can I find one?"

He started to laugh and said, "I know you can't hurt them since you are only a kid. You must be their messenger?"

I didn't know how else to respond, and this seemed like as good a reason as any for me to be on the mountain alone at nightfall without any gear, my pockets stuffed with stolen apples. So I lied. "Yes."

It was not as unlikely at that time as it might seem now. It was hard for the Peshmerga to communicate with one another. Even the technology of the 1980s was not there for them. There was only a small fraction of fighters seeking to gain independence for their homeland. Their support from the outside world was minimal and they did not have proper radio communication equipment at all fronts. This meant that the only way for them to know what was going on was messages in writing, signed by other unit captains, reporting on daily operations and news of the front: who was injured, who was killed, how many prisoners they had, where they had attacked or had been attacked, what supplies they had gathered, etc.

The shepherd shook my hand and said, "My name is Rojhat, I am from Baze Village." *Hey*, I said in my mind. *He is from Oshane's village*.

"Nice to meet you, Rojhat. One of my uncles is named Rojhat."

He smiled. "Is he a shepherd too?"

I looked at him, laughed, and said, "Yes, sometimes."

We both laughed and he continued, "The Peshmerga units are nearby. Their bunkers are everywhere, but I wouldn't go tonight if I were you. The closest one is about an hour walk. If you try to go tonight, they would probably shoot you because they wouldn't know who you are."

Uh oh. Perhaps he did not believe that I was their messenger? Or perhaps he believed I was a messenger to them from another Peshmerga unit. That is what I would tell him if he asked. For now, I kept my mouth shut and nodded.

Rojhat then said, "I will get a fire ready. Find a few pieces of wood." He handed me a hatchet to use for the trees.

I walked to a few scraggly trees and

started taking dry branches off them with the hatchet. I collected two full bundles of wood and laid it by where Rojhat had made a fire pit.

He waved politely at a spot on the grass near him. "Sit down."

I did.

He had lots of supplies. He set food out for us: Kurdish bread, tahini, mast (Kurdish yogurt) and jaji (Kurdish cheese). He pulled out a small teapot and the kettle of water from the fire, and poured enough tea for both of us. I told myself, *I was saved by a shepherd*, saying it over and over in my head. Just like my name, Shivan, which meant shepherd in Kurdish.

Kurdish shepherds were good at what they did. They would be out among the mountains for weeks, if not months, raising sheep and other livestock to feed their families. Most shepherds worked for someone else who owned the livestock, which were mostly sheep. Some people had money, but not the time to raise the sheep. So they hired shepherds. It was a job, though, and many people became shepherds. In some ways it was the best job to have, living around a remote village.

Rojhat the Shivan treated his guest Shivan very well. He fed me and let me stay with him overnight by the fire pit. He wanted to talk, and I listened. I did not respond much. I did not want to give myself away. If I did, the next day he would probably send word to tell my family where I was. My father was probably already looking for me. My dream of becoming a Peshmerga would never happen and I would be sent to Zakho to go to school again.

Rojhat asked if I was ready for bed. I was tired, but wondered how we were going to sleep. There was no bedding laid out, no pillow or blanket. He tossed me his provisions bag and said, "Use that for a pillow." He gathered his long jacket, rolled it up, and used that for his pillow. He simply lay down on the stony ground facing the fire, relaxed with a sigh, and seemed to fall asleep at once.

I laid there on my back by the fire near him, on top of rocks that were stabbing me all over. It was tough, but I just looked up into the sky. Staring at the stars, wondering whether Peshmerga also slept on the ground like shepherds, and what they did when it rained or snowed, my eyelids drooped, and I fell asleep by the warm fire.

The next morning, Rojhat woke me up early. Once again, he took out some items for breakfast, and I contributed my last yellow cherries and an apple. We ate, drank a tea, and he even offered me a cigarette which I started to smoke with him. He had rolled two cigarettes using homegrown tobacco.

We both got up and said goodbye to each other. Packing up his simple gear was the work of a moment; I had nothing to pack. I thanked him for his hospitality and company, and he wished me good travels. He led his flock and dogs west, towards my village, and I went east on Metin Mountain, toward the Peshmerga bases Rojhat pointed me towards.

I had walked for about an hour, just as Rojhat had said, when I spotted the Peshmerga. There were different units camped in the bunkers, though. Some of the bunkers had women with them. The colors of the flags represented their party flags. As I drew near, I realized the different units were all grouped into just two factions.

I saw the flags of Kurdish Democratic Party (KDP) flying high with more than one hundred Peshmerga camped in and around various bunkers that were next to each other. Their flags were all yellow with "KDP" written in red in the middle.

The faction encamped near them was Komala, or the Kurdistan Socialist Party. They had women fighters with them. Their flag colors were different. Some of their flags were red with a Socialist symbol in Kurdish and Arabic script. Other Komala flags were blue with the famous Kurdish flowers, nergiz, on them. There were about thirty to fifty Peshmerga in those bunkers.

Both parties were camped about five hundred feet apart from one another. I had to decide which bunker I would visit first.

It was a no-brainer: I started walking towards the KDP bunker.

Here the Peshmerga would be sure to welcome me because most of them would know my father. Besides, we were in Berwari Bala, which meant most of us would be from the same tribe, and they would recognize that too. I would be welcomed with open arms. I felt confident that they would accept me as one of them. I strode confidently towards the Peshmerga of the KDP, feeling happy and nervous at the same time.

CHAPTER 3

Becoming a Freedom Fighter

I reached the nearest KDP dugout. It was about three feet deep and topped with a foot-high wall along the edge of the entire dugout, made of natural rocks fitted together without any mortar or concrete. Trees covered their area where they could blend in to camouflage themselves. There were men of all ages in the KDP, some as old as their late sixties, and some as young as eighteen. There weren't any teenagers as young as me, though, as far as I knew. I looked around the bunkers at the base and the other dugouts to see if anyone here was around my age. It became clear that I was the youngest one there.

They did not aim any rifles at me as I peered about. They probably thought I was lost or was going to ask them for something; maybe cigarettes, not the rolled-up kind but the ones like they smoke in the city, that come in cigarette packs made at the factories.

I said, "Silav, che hale hewe ye" in Kurdish, which meant "Hello, how are you guys doing?"

"Ser chava, bixer hati," they responded (Greetings, and a warm welcome). Some of them asked me to sit down. I found a place by the dugouts and sat.

The KDP Peshmerga started to pepper me with questions: Who are you? Where are you coming from? What can we do for you?

I told them my real name, what village I was from, and told them I just wanted to stop by and say hello. They were satisfied with these answers and seemed happy to see me.

We started to talk about other things. However, it bothered me that I did not recognize any of these men. Maybe they knew my father, but I did not know them. I did not tell them whose son I was yet. I thought it was too early to say anything about my dad. Most likely they would know him and send him a message saying I was with them. Perhaps it was best that I did not recognize them; perhaps they did not know my family either and would not report my presence to my father.

But if their leader could find no one in the unit to vouch for me, would they let me join their unit? Did their leader know my father?

With this in mind, I asked directly who their commanding officer was. They looked at me strangely and I felt like a fool for not asking more discreetly. It had sounded like a question from an interrogation, when Peshmerga were captured by Iraqi soldiers.

The Peshmerga were very sensitive about talking to people they were unfamiliar with, Kurdish or not. The did not want to leak any information. Their struggle was hard already. They had lost many commanders because of Kurdish traitors, the ones who joined the Iraqi army and mentioned the commanders' names to the Iraqi regime.

Then the regime did not need to hunt down the commanders directly, they would simply arrest the commanders' families. This was leverage for the Iraqi interrogators to force the Peshmerga commanders to surrender themselves. Many of them had no choice but to do so, for the sake of the safety of their family. If the Peshmerga commander refused to surrender, their families were relocated to Arab cities south of Baghdad. When the Peshmerga fighters or commanders were captured or surrendered voluntarily, many of them were imprisoned for years and tortured. Rumors had it many were sent to the war with Iran. Others who held high ranks in the Peshmerga, or those who had strong influence amongst the Kurdish people, were executed right away.

I felt weird for a second that I had asked that question. I told myself in my head, *why did you ask that of them?* Some of them started walking away from me. I had to tell them what I wanted, otherwise I might be arrested on the spot. My village was far from them and they might suspect that I was a spy for the Iraqi troops. I stammered, "I want to know who your commander is because I want to become a Peshmerga."

Some of them looked at me and started laughing. Others looked at me seriously but proclaimed that I wasn't the right age to fight.

In my mind, I knew I was not too young to be a Peshmerga. I tried to explain to them that it was a good choice for me, because it was better to be free than be under a foreign government forced to learn their language in school. I told them that I had been walking for two days to reach them. I wanted to help the Peshmerga and be part of the cause, to join them in freeing our people. Those who had laughed looked more carefully at me now, but those who had frowned at my young age now turned their faces away, but not in anger.

I saw a man appear in the distance walking towards me. He was a tall, heavy man and had a mustache. Several Peshmerga walked along on both sides of him. I assumed that they were his bodyguards.

He approached me and stated, "My name is Sebri Guli. I am the commander of this unit. I know who your father is, and you are too young to be a Peshmerga. As a matter of fact, you already are a

Peshmerga but just aren't ready to fight yet. You still have a few more years left, my nephew."

"Please sir, I am not too young. I can fight. I want to fight. Let me join your unit. I don't want money. I am here as a volunteer. I just want to be a Peshmerga."

He folded his arms over his chest. "Listen, at dawn tomorrow some of my Peshmerga will take you back home. But now, come eat lunch and spend the rest of the day with us. We have plenty of cigarettes and tea. Don't stay in the open because of the Iraqi jets and helicopters."

I could not believe what he was saying to me. After everything I had gone through, Commander Sebri was ignoring me. He didn't accept me as a Peshmerga. Frustrated, I wanted to run away from them but did not want to show my weak side. I had to act like a man and take it in. This was normal.

In the Kurdish culture, older men always knew better because they were the wise ones. They had gray hair, a symbol of their wisdom and experience. Respect was a big thing. I could not talk back. The only thing I could say was "Bele seyda" (yes sir).

I waited until close to sunset just before it got dark outside. After telling them I had to use the bathroom, I slowly started walking away from them, farther and farther away until I knew they could no longer see me. I headed straight for the Socialist Party bunker. I could see the colors of their flags flying. Raising my hands, I walked towards them.

I could see five women with their AK-47s pointing at me. Some yelled, "Raste!" while others shouted, "Boste!" (in the Kurdish Sorani dialect). Both words meant "stop," of course. I slowly took baby steps.

Two men jumped from their bunkers and ran toward me. Luckily, they did not point anything at me even though they did have their rifles and other ammunition with them as they approached. One of them was from our area and the other spoke in the Sorani dialect, which meant he probably came from Hewler, Sulimaniya, or Kerkuk. They asked what I was doing at their base. I responded that I wanted to join them and become a Peshmerga. They smiled and patted me on the shoulder. With each of them by my side holding my hands, they took me into their base. They held my hand out of care and respect, out of love and feelings of fellowship with me, their countryman. They kept on saying, "Bixer hati, bixer hati" (welcome, welcome).

The men took me into their bunker area and introduced themselves. Their commander was named Hishyar. He was in his late thirties and looked like an educated man. He was six feet tall, looked like he needed a haircut, and wore glasses. He talked about the struggle of the Kurdish people saying, "We must all be fighters: men, women, children. This is how we will gain our freedom." Hishyar believed that everyone in Kurdistan should be equal, that no one should be poor and that the government's job was to make sure that everyone shared the resources in the land.

Finally Hishyar asked, "Can you use a rifle?"

I replied, "Yes."

He called to Shirin, a woman Peshmerga. "Shirin, can you grab an AK-47 for Shivan?"

Shirin replied, "Bele seyda" (yes sir). She brought my weapon, and it was a Russian AK-47. It had one magazine in it. Hishyar handed me two more magazines along with a belt that could hold grenades and other items. I was also given a backpack.

I told him I didn't know how to thank them. In my mind, I couldn't believe what was happening. Was this all a joke or truly real? "I won't let you down," I told them.

Hishyar looked at me, smiled and replied, "Good, because at first light we are going to fight."

I finally became a Peshmerga just like that. On my second day I was going to be in battle, and I was only fourteen.

Hishyar's unit consisted of fifty Peshmerga. I was the fifty-first in their unit. They had about forty men and ten women. Most of the women were young and from the regions of Hewler, Suleymaniya, and Kerkuk. They all spoke the Sorani Kurdish dialect. They were impressed by my bravery and spoke well of me to Hishyar. In my mind, I was also impressed by them, that they were women and that they had left their families and the comfort of being a woman to fight for a great cause.

I did not talk about pay or rations of

food. I felt good holding my rifle. Being around the Peshmerga felt good. I was finally fighting for my country. I was going to be doing more in whatever way I could.

That night, my mind was spinning, thinking about Beri, my mother, and my siblings. I did not really think about my father. I knew he didn't approve of what I was doing. The Peshmerga unit I was with had not asked for my age, and I didn't care to tell them how old I really was. They had given me a weapon, and I was going to use it on my enemies. I asked myself, *What if I were to get killed in the fight? Who would know where I am? How would my unit take me back to my village?* More questions and thoughts were coming to mind. I was staring into the dark, moonless night with only stars in the sky. It was a bit chilly; the hair on my arms stood on end. I slowly got closer to one of the fire pits. Lying next to the bunkers, my eyes slowly drifted shut, and I went to sleep.

Morning came, and I opened my eyes slowly. Suddenly I heard, "Bine rez" (line up). Everyone hurried to form their line. I stumbled around at first, still half-awake, but managed to tag along with the unit slowly. I got into the line with my rifle and other gear that I had been given by the Socialist Peshmerga. Hishyar made a speech:

"Today we are here because we choose to be. We are here because we are fighting for a cause that will free our people. We are doing it in the name of the Kurdish people, in the name of our homeland, Kurdistan.

"We did not start this fight; we are fighting for our rights. We fight to live in a free nation where no Arab, Persian, Turk, or any other foreigner will rule us. We are the children of the Meds, the children of Kawa the Blacksmith, symbol of Newroz. Today is a great day because we will fight for our honor. We will not surrender, and we will take what is ours, our homeland."

Everyone started shouting.

Goosebumps ran over my body. This was real. This is what I always wanted. I screamed as loud as I could, lifted my rifle into the air, and shouted, "Yah!"

Hishyar pumped his fist in solidarity and turned away. Shirin took his spot and called out, "Okay, we will eat and then we will move out. Make sure you clear everything out of your bunkers." We all got to work while breakfast was prepared.

Everyone did their part: gathering wood, starting the fire, putting the tea kettles on the fire pit, cracking eggs and putting them on the pan, peeling potatoes, chopping tomatoes and other vegetables into the pots on the fire pits. There was bread available so there was no need for baking.

It took us about two hours to get breakfast ready and to finish eating. Then everyone packed and got into a line, and we started heading south over the top of the Metin Mountain. We were heading towards Duhok. That is where most of the Iraqi military bases were. We walked for about two and a half hours when Hishyar said, "Okay we are nearing that base, which we will attack." The base had Iraqi soldiers in it. Even though we were still at a distance, they could not spot us because we started crawling and hiding behind rocks and trees. We could spot them though. We were at a higher vantage point on Metin Mountain. Shirin also had a few binoculars that were shared by the Peshmerga on our unit.

I could see the Iraqi machine gun muzzles poking out of their bunkers. I didn't know how many soldiers were in it exactly, but I knew we had more men than they did.

Shirin whispered, "Alright! Ready?" She then raised her hand and called out, "Hejrum!" (attack!).

Shirin started running towards the Iraqi soldiers' bunker along with the other women, the rest of the Peshmerga right beside them. I readied my rifle and started to run right behind them. The first thing I was told was, "After the first shot, take cover." As I was running with the rest of the Peshmerga towards the bunker, my eyes wandered around, looking for big rocks to shield myself from the bullet fire that was about to come. We started firing and hell broke out.

The Iraqi soldiers started firing back with machine guns, AK-47s, and RPGs. We got closer and closer to their bunker. A couple of our Peshmerga were hit while running to their bunker. I could see them on the ground, some of them in a lot of pain. I took cover behind a rock close to the Iraqi soldiers' bunker. Another Peshmerga came by me, and called quietly, "Do you have grenades?"

I said, "Yes," and handed him a couple.

He took both and started crawling to the bunker. A few minutes later, boom! Followed by another boom! The Peshmerga managed to get both of the grenades into the bunker. I could hear the Iraqi soldiers screaming in Arabic. We then surrounded the bunker. Three Iraqis had died with four others still alive. Two were injured badly, hit by bullets and shrapnel from the grenades. We had lost two Peshmerga, one female and an older man that was in his late fifties. There were ten other Peshmerga from our unit that had been injured, thank God nothing serious though.

Hishyar said, "Grab what you can; it is all yours. The food and heavy weapons are for the unit. Keep pistols and ammunition and other valuables for yourself." I was thinking about the Iraqi soldiers. I wanted to kill them right there. It was finally my chance to get revenge. They looked scared though, weak and calling for mercy in Arabic, "Rahmallah" (for the sake of God).

Shirin said, "Take the soldiers; make sure they are in front of the unit. We need to clear out before the helicopters come."

That day taught me a valuable lesson: life is short. Anyone can die at any minute. I was glad to be alive. A couple of hours earlier the slain Peshmerga woman had been happy, smiling and singing Kurdish folk songs. The older man had been telling us about his life, the struggles he endured raising a family, how he was going to fix his children's future.

I was sad to lose our Peshmerga. It made me think about my family, my village, and my love, Beri. I thought about my mother and what she was going through, probably worried sick and crying. My mother was always sensitive. She was probably thinking that something terrible had happened to me.

I had to go back.

Even though the struggle of my people was important, this battle had opened my eyes to appreciate life. I wanted to see freedom one day for my homeland, but I had to go back to my family and Beri first. This fight would be for another day. I had come close to death and was happy to be alive. I must fix my mistakes with my loved ones and then think afresh about our fight against our oppressors. Maybe there was another way to free my people. Maybe my father was right, that an education was far more important than what I was doing, that a pencil could one day free our people more surely than a gun.

We camped out about an hour from where we attacked the Iraqi soldiers' base. I was thinking of ways to ask Hishyar, my commander, for permission to go home. What would I say? How would I approach him and convince him that I needed to go back home so soon? I was too shy though. I did not know how to approach Hishyar and the other Peshmerga in my unit. I was new and didn't want to show my weak side to them. The best option was to leave my rifle and other weapons provided to me behind. I would start heading back home at first light the next day.

Hishyar asked who wanted to stay up and patrol the area we were going to camp at. I raised my hand and volunteered along with a few other Peshmerga.

My plan was to leave that night.

Night came and it was pitch black. I set my rifle by the campfire, with my pistol and other ammunition. Before dawn the next morning and before any Peshmerga was awake, I left the unit and headed back to my village, Belizan, to be with my family. I walked slowly away from our base for about five minutes. When I knew I had a good distance, I started to run. I ran as fast as I could. I did not know what direction I was heading. I did not look back, though. I just ran, sometimes changing my directions.

CHAPTER 4

Losing My Father

I walked for about two days. As I was neared villages, I asked local villagers where I was and what village I was near or on my way to Belizan. One villager mentioned that I was near Eridina village. I decided to walk there.

My mother mentioned in the past that we had some extended relatives in Eridina village. She told me, "Your grandpa, my father, was married to my mother, your grandma, at the time when my mother died of childbirth, from me. He later married another woman from Mosul who became my stepmother. Her name is Gulbuhar. After your grandpa passed away, she remarried a man named Hejar and moved to Eridina village." While I was walking back home, I remembered this conversation that we had. I remembered the person my mother's stepmother married and remembered that his name was Hejar. He was good friends with the villagers in Belizan and my relatives.

It was getting late, and I didn't want to be stuck in Metin Mountain overnight. I walked towards a few villagers. I greeted them, saying, "Aslameleykum" (peace be upon you in Arabic). They responded back, "Welekumsalam" (peace be upon you also). I asked where Hejar Eredini's house was, and they pointed to a few houses back, saying, "Right there with the kids in front playing."

I responded, "Sopas" (thank you) and walked towards his home in the brightening dawn.

I knocked on Hejar's door. A few kids showed up at the door. I assumed they were his new kids. They hollered for their father and he came. "Bele?" (yes?) he said. He was a light-skinned man of medium height. He was definitely taller than I was. He had a light mustache with black hair.

"My name is Shivan," I told him quietly. "I am Salih and Aysha's son, from Belizan."

He looked at me smiled. "Well, aren't you going to come in?"

Hejar was very happy that I had stopped at his house. He looked at me in awe. "You are all broken, and your clothes are torn!"

Gulbuhar, my step grandma, walked in and grabbed me. Prior to marrying and moving to Eridine, my grandmother was known for her looks. After my mother's father passed away, people from all villages of Berwari Bala came to ask for her hand. Even men from the city came. She was very pretty with long black hair, big dark eyes that glowed when she looked at you, and was about five-and-a half feet in height. She kissed me on both of my cheeks and said, "Bixer hati" (welcome). She then walked into the kitchen and started cooking right away.

Hejar grabbed his tobacco and rolling paper, made a cigarette, and asked me if I wanted one. I took the bag and rolled a cigarette. We smoked our cigarettes while drinking our tea. He looked at me and said, "De bu min beje" (go ahead and tell me).

I looked at him and smiled. "All of it?"

He smiled too, and said, "Keyfa teye" (it's up to you).

I shared my story and the journey I had. Hejar, too, shared some of his own story with me. He looked much older than he really was. He looked torn down by all the years of labor from farming, and from being a Peshmerga too. He was not a Peshmerga on a regular basis but did sympathize with them while he was taking care of his family and his farm fields. He said, "I would love to be in the mountains with the rest of the Peshmerga at all times, but my family needs me." His children were all small.

I said in my mind, *I don't blame you.*

"I can't keep wandering off. I need to take care of my family." Hejar said, "The Iraqi government has bigger weapons and more soldiers. It is impossible for us to beat them alone and under these circumstances."

After I told my story to Hejar, he lectured me on listening to my father, saying, "Since you are the eldest child, one day you will follow your father's footsteps." He continued to say, "One day you're going to have to take care of your family after he is gone." Hejar told me about my mother's worry, that if you hurt them in this world, even if you make them sad, it was impossible to see heaven, according to the Quran. He started saying, "If your mother does not forgive you, you shall not enter paradise." He told me, "Mothers go through a lot: carrying you in their stomach for nine months, changing you when you get dirty, feeding you. Nurturing you as a baby is not an easy task, especially if you have other siblings. On top of that, your mother has to cook and take care of all the household chores."

Hejar and I continued to talk throughout the night.

My step grandmother, Gulbuhar, would tell me stories about me when I was younger in Belizan. She kept laughing. "You did this and that. You were a troublemaker," she kept on saying. My step-grandmother continued to talk about my family and the struggles they endured in Belizan for being from another village.

Later that evening, they had made my bed and it was getting late. I needed to leave their home early in the morning before dawn. The Iraqi soldiers were near their village, stationed in Bamerne, one of the biggest air bases in the Duhok area. I thanked Hejar and Gulbuhar for their hospitality and giving me shelter. I told them that I would leave early in the morning and that they may not be able to see me before I leave. That is why I said goodbye before I went to bed. It was my first time in Eridina. I loved the small village and especially the company of Hejar and my step-grandmother.

I woke up early the next morning. Thinking I was the only one awake, I was surprised to see Hejar outside by the doorway smoking a cigarette and my step grandmother making breakfast. I said "Rojbash" (good morning). They replied back, "Rojbash." We had breakfast and I had to say goodbye again. I wanted to be home before dark.

I excused myself and Hejar told me I was about two hours away from my village, which was walking distance. He gave me a bag that had a bottle of water, some bread, and Kurdish yogurt to take with me. I thanked him and headed home.

I had been walking for several hours. It was midday. I was tired from walking. My shoes looked like they were going to fall off my feet. I kept walking as the sun was turning slowly to the west side. Seeing the beautiful mountains, water streams, birds chirping away helped me to get my mind off of how tired I was. It felt like I was walking forever. Going over mountains, rocks, and bushes is not an easy task.

As I crossed the Metin Mountain and approached the other side coming down, I could see my village and the dogs barking from a distance. Men were working in the fields. I had been gone for about six months. I missed everything. Everything looked so beautiful. I had taken these things for granted. I hated my village before. I always wanted to leave and never come back. When I saw it now, it was the most amazing place that I had ever seen.

I walked through the farm area. Villagers greeted me; some came to me and shook my hand. Some of my friends came running to me. They hugged me and said "Man, you are a qehreman," which means "hero" in Kurdish. I did not want to chat with anyone. I just smiled through most of the conversations. I wanted to go home, hug my mother, kiss my siblings and take a nice bath that would cleanse me from all of the time I had been gone.

I stepped in front of my house. My siblings saw me and came running to me. I hugged them deeply, with kisses all over their cheeks. My sisters ran inside yelling, "Dade, dade, Shivan hat" (mother, mother, Shivan is back). I could hear my mother drop her dishes and come running outside. She saw me, looked at me, and came to me. She grabbed me and held me tight, kissing me all over my face. She was crying. She said, "Sit down, I will get you new clothes."

She went outside to get the water warmed up for me to take a bath. She came back and said, "Go take a bath, and I will get supper ready." I asked where father was. She looked down, like something happened. She said, "Please Shivan, my son. Go take a bath, and we can talk later. Tu ye westay" (you are tired). I told her okay. I was tired and had dirty clothes on. I needed to clean up.

I went to take a bath. It felt like the best bath I had taken. I was in there for about an hour straight. I dried myself up, put my new clothes on, and went to the front porch to eat dinner. I saw that food was ready and sat down. My mother said, "Eat son, you are hungry." I was hungry but wanted to wait for father. I said, "I will wait until father comes back."

My mother said, "Fine then, I will bring you a tea until he comes home. He should be home any minute now. He has been gone for quite a while."

My mother brought me a tea and sat down. I asked again, where father was. She said, "Something happened in the village. A new family came to stay in our village. The woman in that house had accused someone from the village of trying to rape her. She said that he had burned her foot while she was asleep with a cigarette to wake her. She woke up and yelled. The villagers came to her aid. She accused a young man of doing it."

My father had asked her if she could prove which young man had done it. She said, "No, but I think I know it's one specific person who looks at me all the time." Father was defending the young villager as the woman's husband was threatening to kill the young man. Father said that since there was not enough proof that the young man did it, he should not face any threats, and they were going to protect him along with the other villagers.

My mother said that my father was with my uncles and that he had ordered the other villagers to make sure no harm was to come to the young man. As my mother and I were continuing to have our conversation, I saw father walking in.

He saw me, came to me, and hugged me. He asked, "Did you learn your lesson yet?"

I laughed and said, "Yes, father, that is why I am back."

He said, "I am glad you are alive son. You are too young to be fighting."

I was relieved that my father did not yell at me. I thought he was going to get really upset and give me a beating. He was calm and something inside him gave him relief that I was back home. He sat down and called on my mother and other siblings to come eat.

We sat down together as a whole family. It was the most amazing moment in my life. We were all smiling, laughing and enjoying our meal together as a family. I looked at my mother. She looked happy but was also looking worried. She must have been thinking about the situation involving the woman and the young man accused of rape.

My father was busy eating and encouraged my younger siblings to eat. He was keeping a close eye on them making sure that they did not sleep hungry. My siblings looked happy not knowing the troubles of life, the things that were happening around them. They were just kids in their own little world.

For myself, I was happy to be home with my family. I missed my mother and siblings dearly. I had been gone for six months. I had not agreed with my father on many things and thought he hadn't allowed me to seek the things that I wanted. Now, I knew.

I knew that my father cared for me; he loved me dearly and wanted the best for me. That was why I told myself that

I would never let my father down again, that I would not disobey him now or in the future, that I would listen to him.

We finished eating, and some of my siblings were asleep. It was getting dark out, and in a few hours, we would go to bed. We had to sleep early when there was no company over. We had no electricity and relied on firewood and kerosene lamps to give us light.

My father was listening to the radio. There was a Kurdish Radio Station in Armenia that broadcasted events in the Kurdistan Region of Iraq. He listened to it every night for news and events that he had not heard yet. We were drinking tea in the living room. My mother was in the other room, putting my siblings to sleep, when we heard a noise in the front yard: footsteps followed by a knock. It was my father's friend Zinar from Keste village, about two hours away walking. He came to visit my father. Zinar was a handsome man, about five feet nine inches in height with a full head of hair and a mustache similar to my father's. He wore a nice set of Kurdish clothing, khaks that were dark brown in color.

My father randomly had visitors. We

did not have phones, so people couldn't call and say they were coming over. People relied on foot, horses and donkeys to travel. Cars were very rare in the villages unless someone was in the Iraqi military; not just a regular soldier, but someone with a high rank in the military. Everyone kind of expected someone to show up at any time.

My father was happy to see Zinar. He told me, "Tell your mother to get tea ready and something to eat for Zinar." I took Zinar's coat. It was beaten down and torn in many places, unlike his khaks he was wearing, though it looked very warm. I also took his rifle and put it in the closet as I knew he was going to spend the night.

I helped my mother to bring some food for Zinar and tea. He and my father went out to the porch. I followed my mother into our storage unit where she grabbed a blanket, pillow, and a dushek (small bed) for Zinar to sleep on. I brought it into the living room where he was going to sleep. I set it in the corner and went to join my father and Zinar.

They talked for hours, sometimes in a serious tone when it came to politics.

They did not argue with one another but had deep conversations about the future of the Kurds in Iraq. They laughed at each other, bringing some old memories into their conversation.

Zinar asked my father if he wanted to go hunting the next day. It was fall and mountain goats were popular to hunt. It was mountain goat season, and its meat was priceless. My father had shot a couple in the past, and the meat is by far the best. He also brought us a lot of partridges, or kew, as we call them. It's a popular bird in Kurdistan where its meat is a cuisine.

My father agreed to go with Zinar the next day. He told me to get his rifle. He wanted to check it to see if it was clean. I brought it to him. I also had a surprise for my father. I had taken a pistol and several magazines full of bullets from one of the Iraqi soldiers that had been killed. I wanted to keep the gun because it was really clean and in good shape, but this was going to be my gift to my father.

I went to my room and brought it. I listened into the conversation once again because I did not want to be rude and interrupt. My father told me to go get another tea for them and grab his bag of tobacco so they could roll some cigarettes.

Before I got up, I said "Father, I have something for you."

He looked at me, confused. He probably thought I was going to embarrass him in front of his friend Zinar. I unwrapped the cloth on the pistol and handed to him, with the three extra magazines full of bullets. "I want you to have this."

He looked at me, grabbed my hand, brought me close to him, and kissed my forehead.

I excused myself and went to make some more tea and grabbed his bag of tobacco with his rolling paper. He handed the gun to Zinar, and they examined it. I knew they both were professionals when it came to weapons. You see, when you are a villager, one has a lot of time to study things. Having weapons is also mandatory. There were a lot of wild animals near our village: wolves, bears, wild hogs, mountain lions, etc. Having a weapon was also important if you ran into Iraqi soldiers. Almost every villager that was even a little well off had at least a rifle or shotgun in their house. If they were a Peshmerga, they had a lot more weapons than the average villager.

When I woke the next morning, my father called me. He and Zinar were already awake. They had already eaten breakfast, were drinking a tea and smoking cigarettes. He said, "Son, you don't have to go to school in Zakho or Duhok anymore. They have a school now in Babire." That was about an hour of walking from our village. He said, "It is Thursday now. I want you to start on Saturday. There is a teacher there named Ibrahim. Tell him that I sent you if I don't get back before Saturday."

I agreed with my father. I wouldn't let him down this time. I told him I would start on Saturday.

Later, he and Zinar headed for the Metin Mountain. They looked sharp. If an Iraqi soldier would have seen them, they would mistake them for Peshmerga fighters ready to attack. They both had their khaks on and Kurdish turban wrapped tightly around their heads. Their rifles were on their shoulders. They both had sacks that held their drinks, food and camp gear. My father's pistol that I had given him was in his sheetik, the cloth wrapped around the Kurdish khak trousers and vest.

After my father and Zinar had left, my uncles came to my house. They asked where my father was. My mother told them, "He and Zinar went hunting."

Uncle Rojhat was the talker. He said, "Yes, good for him. We have a million things to solve, and there he goes hunting."

Azad said, "Just let them be. We will talk when he gets back."

Chekdar was the innocent of them. He didn't say much but just liked to repeat what others said most of the time. Or he would make strange comments. He said, "Yah, he shouldn't have gone. We got work to do in the field."

Rojhat looked at him and shook his head. "What are you talking about? We have matters far more important to solve. We need to do something about the young villager being threatened."

Azad raised his voice, "No one is being threatened and no one can do anything as long as I am alive."

My mother brought them tea. They smoked a cigarette. After their tea was finished, they excused themselves. My mother insisted that they stay for another tea. They told her, "We must go to the mosque and talk with some of the other villagers about this matter." I asked if I could go along. At first, they looked hesitant but then agreed, so I tagged along.

Walking with them, I saw Beri on the way. She was at the water well right next to the mosque. My Uncles entered the mosque. I told them, "I will be right in." First, I had to wash my face.

I walked to the well and greeted Beri. She looked beautiful as always. Beri said, "You came back?" I said, "Yes. I got tired and home sick. I missed everything, especially you, Beri." I knew I would see her somewhere in the village alone. I did not want to make a scene. I grabbed into my pocket in which I had brought her a gift. It was from Shirin, the brave Kurdish fighter in my Peshmerga unit. It was a bracelet that she had made, with the colors of the Kurdish flag: red, white, green, and a twenty-one-sun ray emblem in the middle.

I handed it to her and said, this is for you. Beri didn't want to accept it at first. She said, "I am mad at you. You have been gone for too long." I told her, "Take it, please. I don't want anyone to think anything." I handed it to her one more time. She quickly grabbed it, and I left for the mosque. I entered and greeted everyone there: my three uncles, the young man who was accused, his father, and his uncles. They talked for hours.

When it was time to pray, they did, even though Uncle Rojhat didn't. He just sat there and talked. When others were praying in a line, he went out for a cigarette.

It was coming to be near evening, getting a little dark outside. In the middle of the conversation my uncles and the young man were having with the young man's father and uncles, we heard gun shots coming from a distance, followed by more rounds of shots coming from Metin Mountain.

Azad said, "That was quick. I can't believe they found mountain goats already." Then a few minutes later, we heard more gunshots coming from Metin Mountain.

They all got up and said, "Something is not right. Hurry and get ready to go up." As we went outside, everyone in the village was outside. They knew something had happened. Azad called to his wife loudly, "Grab my rifle!"

Rojhat ran home to get his. Chekdar just stood there not knowing what to do, waiting for Azad and Rojhat's orders. Some of the other villagers came too, with kerosene lights and guns. My mother came out screaming. She kept on saying, "Please go, please go. Something is not right. Please go." I hurried to my house and grabbed my father's extra rifle. There were about twenty of us who headed for Metin Mountain to find out what was going on. Some of us had rifles, some had kerosene lamps, some had sharp blades that cut the tall grass and others just came along to help.

As we were walking, it got darker and darker. We could barely see each other, except for those of us next to the light. We formed a line and followed each other's path. The walk seemed forever. We walked for over three hours, and we could barely see anything anymore. We yelled for my father and Zinar. No one replied back, no response.

Everyone was worried sick. They kept on saying, "Something is not right. Something has happened." The group was tired after walking for hours. It was around 10 PM. We all stopped and rested for a little bit. My Uncle Rojhat suggested that we disperse into smaller groups and go in different directions. We had enough kerosene light to do that. This would help find them faster. Three of the villagers with us went back home.

Azad formed a group of five, including himself. My Uncle Rojhat formed a group of four, including myself and two other villagers. The rest of the villagers went their way, and Checkdar followed them. We kept on calling my father's name, "Salih, Salih." No response from him. We walked for about another hour and came upon Zinar. He came rushing to us. He could barely walk. He said, "He's hit! He's hit."

My Uncle Rojhat grabbed him by the vest and said, "Tell me right now: what is going on?"

Zinar, gasping for air said, "We got ambushed. We got ambushed." My Uncle Rojhat asked Zinar, "Which way?"

"Up," Zinar replied, "Up."

We walked up for about ten minutes and there I saw a body. I ran to it. I slipped but my hand touched it and got wet. It was my father's blood and body lying there, right on a tree. The tree managed to stop his body from rolling down the steep mountain.

The light came and I looked at my hand. It was full of blood. I did not know what to say or do. I was in total shock. I just let it out. I let it out as hard and as much as I could. I cried and cried. My father was there next to me, dead.

One of the villagers gave out the signal that we had found them, one shot in the air. One shot came from the second group, followed by a third shot from the last group. It was over. We had found my father but not in a way that we had hoped for. Zinar grabbed my father's side. My Uncle Rojhat grabbed the other. The other two villagers grabbed his feet. I grabbed his head, which felt very strange. He was shot in the head.

I could barely hold my tears and keep my strength walking back to the village. The other groups caught up to us. Some of them were crying out loud. My Uncle Checkdar held my father and didn't let him go. He said, "No, no. Who did this? Who?"

We finally made it back to the village

after a long walk. It was midnight. It seemed like day though. All the villagers were still awake, even the children, wanting to know what had happened.

My mother came running, "Where are they? Where are they? Where is Salih? Is everything alright?" My mother continued to repeat her questions. When she saw my father's body, she just collapsed. Other women from the village went to her rescue, and we took my father into the mosque.

That night was the worst one of my life. I did not get one minute of sleep along with some of the others in the village who were inside and outside of the mosque. Morning finally came, and we saw the devastation of my father. He was shot in the head, ribs, and arm. Part of his head had been shattered from the bullets.

We were going to bury my father. My Uncle Azad suggested we wait until my grandpa arrived. He had been traveling to Duhok for some work. Azad would go and get him. Some of us were inside of the mosque in silence, some were reading the Quran, and others were outside of the mosque in total disbelief.

Azad and my grandpa arrived later that day with about ten other men. Those men heard about the news and wanted to come along for the burial. The news reached all the other villages near Belizan in Berwari Bala area, including Brifka, where most of my relatives were.

My father was not just a regular man in the village. He was well respected. His status was a Peshmerga commander under cover. He liked people and was always finding ways for people to come together. He did not like problems and always found a solution for disputes between the villagers.

It was nearing late afternoon. There were over a thousand people who came to my father's burial. The population of Belizan was about two hundred. The villagers had never seen so many people at a gathering like my father's funeral. I felt really bad for my mother. I was feeling her pain. She was nowhere to be seen. There were so many people that it was hard for us to keep track of who was where, especially the women of the village and others who came that were comforting my mother.

An hour later, the mullah announced

on the mosque speakers that it was time to head to the graveyard. My father's body was washed up and put into a white sheet of cloth wrapped around him. He was born naked and was going to be given back to the earth the same way, except for the piece of cloth around him for privacy. This was the Islamic way to bury the dead.

One man after another was carrying my father's body. I was getting closer to his body. Every time I reached for him to help carry his body, I lost my strength; I could not do it. So, I just walked, sometimes stumbling and other villagers holding my hand and having their arms wrapped around my shoulders for support and comfort.

We arrived at the graveyard. My father's grave had already been dug by Uncle Chekdar and other villagers. They lay his body by the grave. The mullah of the village stepped to the front. He did the prayer for the deceased with others standing beside and behind him.

I shut my eyes and talked to God. I asked him, "*Please forgive my father. He was a good man. Grant him the paradise that he deserves.*" I was not religious at all and never had been, even though I knew the foundation of the Islamic religion from school and home. I truly prayed that day though and knew that God was listening.

When the prayer was done, my uncles and other villagers placed my father's body into the grave. I was the first one to be given the shovel. I grabbed some from the ground. I hesitated for a minute and threw the dirt on his body. I then dropped to the ground, bursting into tears. I saw arms wrap around me from all sides. They took me away from the gravesite. I was screaming in tears as family, friends, and the villagers walked me away. Losing my father was the toughest thing that I had seen in my life.

CHAPTER 5

Berivan Gets a Marriage Proposal

A month had passed. Things were not the same anymore. My father had left an eerie feeling in the village. My uncles were not getting along with the other villagers.

You see my grandfather, Reshid, came from another village prior to my father and uncles' birth. He met my grandmother, Fatima, who was a widow. My grandmother was twenty years older than him. He worked on her lands. My grandpa kept her farm up the way she wanted it, and when it was time for harvest, there was plenty of everything. My grandma saw how hard a worker my grandfather was. She was impressed by him. Soon, they fell in love.

My grandfather was a poor man and didn't have much. He had half-brothers from a different mother, but same father, and they didn't get along very well. He came to work for my grandmother in Belizan. They got married, and he stayed. All my grandmother's land, of which she had a vast amount, was given to him after their marriage. Then, they had kids but the villagers of Belizan always looked at them as outsiders. Even though my uncles and father were all born in Belizan, they were considered Brifki. Now that my father was gone, it was going to be tough keeping things peaceful between my uncles and the other villagers. My father was a peacemaker, but he was not around to keep the peace.

I was working in the fields on a hot summer day when I got thirsty and tired. I decided to stop and go to the water stream to cool off and get hydrated. You see, the water stream is very pure and clean making it safe to drink. We used the water for everything. I came to the stream and saw Beri on the fields. She was picking apples and other fruit from the orchard trees. I stared at her, trying to make eye contact. She was too busy.

After trying for a few minutes, I had no luck, so I rested by the stream and laid down after a while. Laying on the ground felt good. I stared into the beautiful baby-blue sky. The few clouds above me made shapes of creatures and other things. The sound of water coming from the stream, birds chirping away, dogs barking near the village, sheep and donkeys making their noises and villagers having conversations gave me a sense of peace.

Beri's voice came to me, and I looked up. She said, "You are a hard worker."

I leaned halfway up on my elbows. "Yah, I take after my father."

She said, "I am really sorry. I wish that never happened. My father is very sad and talks about your dad all the time. He says, 'Why do the good always die young?'"

I thanked her, and she gave me some fruit to eat. I washed an apple in the stream next to me and started eating it while continuing our conversation. I told Beri, "This apple is sweet just like you."

She looked at me with a beautiful

smile that revealed her cute dimples and started to laugh. Beri and I talked until we heard other villagers walking towards us. She said, "Listen. I have to go." She looked at me and said, "I love you, Shivan." I didn't know what to say; my mind had gone blank. I just smiled at her. As she walked away, I called her by a tender nickname, "Beri, Beri!"

She looked back and I said, "I love you too, more than anything, Beri."

She smiled and put her finger by her lips to say, "Be quiet." She didn't want anyone to hear me. I didn't care though. I really loved Beri. I was going to marry her.

Now that my mother was alone and my siblings were young, I needed to get married to have someone to help my mother around the house. I needed to start working to start making money to take care of the family. I didn't have the connections like my father did. He had support by the Peshmerga. When they came, they would bring him supplies and vice versa. He had a store and had people who gave him supplies from the city. I didn't have these connections, nor did I know how to connect with them or who they were. I had a choice: to farm and support the family with food only or join the Peshmerga once again for a small salary.

A week had gone by, and I saw some older men come into the village that I had not seen before. They were well dressed in Kurdish bergiz, top-of-the-line Kurdish clothing. They all had sacks of items with them. There was also a young man a few years older than me with them. I automatically knew what these men were about: they had come to ask for the hand of a bride. Who was it though? There were not many young women in the village. Well, there were but only a few good-looking ones. These men that presented themselves obviously were looking for a beautiful bride that they knew was already in the village. They obviously had some kind of connections, and someone had invited them. They wouldn't just show up and ask for a bride. They were invited by the father of the bride-to-be.

When it comes to marriage in the Kurdish culture, a man has to like a girl. The girl has to know that the man likes her. Then the man asks his parents to talk to the girl's parents for permission to introduce themselves. They go and give them gifts, followed by a marriage proposal. If the parents of the bride agree to give their daughter to their son, then they plan and celebrate that day and night. It is followed by an engagement with the man's family coming to the bride's village later that year. After a wedding date is set, the man and his family come back to pick up the bride to take her into his village for the actual wedding ceremony.

I was not going to sleep very well that night. I wanted to know who the bride was going to be. I had a bad feeling, but I told myself, impossible. She would never do that to me. She said she loves me. I told myself it's all a mistake, it must be someone else. I finally went to bed after the kerosene lamp went out on its own.

The next day, I went into the fields to work. It was nearing midday when I heard gun shots. Gun shots can be good or bad. These gun shots sounded like excitement shots coming from the village. I could hear people laughing when the gun shots went off, and in bad situations, people don't laugh.

I dropped my shovel and headed for

the village. I came home and washed up. My mother came and asked if I was hungry. I said, "Yes." I asked her what the fuss of the gun shots was all about. She said, "You know the men who came to the village yesterday?" I said, "Yes. Okay, go on." She said, "They have asked Beri's father for her hand. He has agreed."

My head started to spin; I was feeling dizzy. "What? Are you playing a game, mother? This is not funny!"

She said, "What? Are you okay son?"

"No. I am not okay. I am the one in love with Beri. She is mine. We were supposed to get married."

My mother said, "Honey, you are too young. There will be plenty of girls for you to choose in your life to marry."

"No. I don't want to marry anyone else. I want to marry Beri. I wanted to make her my wife as soon as possible. I wanted Beri to come into the family to help you around the house...."

My mother interrupted my rant saying, "Son please, don't do anything crazy. Beri belongs to her parents first. They know what's best for her. Leave them alone. If things don't work out, then we can talk about it."

I didn't smoke much before, but I wanted a cigarette badly. My mother brought me food, but I was already full. Full of anger.

I saw some young guys my age by the mosque smoking. I went to them and asked for a cigarette. They gave me one and lit it for me. I sat down with them and smoked my cigarette.

My Uncle Rojhat walked past me. He was in full gear, looking like a Peshmerga fighter. He had his rifle, a pistol, several bullets wrapped around him, and a backpack.

I ran to him . "Where are you going?"

"Nowhere. I am going to the mountain."

I did not believe him. He looked like he was going after someone, like he was going into battle.

I asked him again. "Uncle, please tell me, where you are going?"

He said, "Shivan, we have received good news. We know who killed your father."

I looked at him, goosebumps running over my body. "Tell me more," I encouraged him. "You have to tell me; it is my father who is being avenged."

He said, "You know about the incident that happened in the village? The accusation of rape? Well, that lady's husband joined the Iraqi army. He brought soldiers with him saying that he knew of a Peshmerga commander, which was your father. He wanted the Iraqi intelligence agency to arrest your father, and this was an easy plot for him to get his revenge on the young man in the village. While the Iraqi soldiers would arrest your father, he would go into the village and kill the young man. However, your father and Zinar found the Iraqi soldiers. Your father confronted and arrested the woman's husband for treason. This is when the Iraqi soldiers killed him in an ambush. I will now go get your father's revenge."

I told him, that I was going with him. Of course he said, "No. It is too risky." I did not listen. I went inside my house quickly. I did not pack anything. I just grabbed my rifle and the pistol that I had given to my father. By the time I came outside, I could see my uncle was way ahead of me. I ran as fast as I could to catch up to him. I finally caught up.

Uncle Rojhat said again, "Please,

Shivan, go back. You mother and siblings need you. Both of us cannot avenge your father. This is a secret mission, and us being together will make the Iraqi soldiers suspicious."

Word had come to my uncles about where the Iraqi troops were stationed and the lady's husband with them. They were in Duhok. Duhok city was full of Iraqi troops, police, and secret service. Anything that looked suspicious was easily dealt with. I did not listen to my uncle. I kept on tagging along.

We came close to Babire Village where I was studying. It was getting near dark; my uncle knew of a family there, and so we went to their house for the night. Morning came and my uncle was nowhere to be found. I asked the man of the house where my uncle was. He told me that he had left before light. That his orders were to tell me to go back to the village.

I did not know where to go to avenge my father's killing. I did not even know how to get around in Duhok, because there were so many Iraqi troop checkpoints. I had no choice but to head back to the village.

After I had breakfast in Babire, I left for Belizan. It was around eight in the morning. I would get home by noon.

I was very upset. Losing my love with Beri getting a marriage proposal and then not being able to avenge my father were going to be memories that I would have to deal with for the rest of my life.

On my way home, I saw a few kews, (partridges). I hid by some rocks and shot at them. I managed to kill a couple. I took them with me to the village. I made it home around noon.

I handed my mother the partridge birds to cook for lunch. My mother asked, “Where did you go?”

I did not want to tell her what was happening, what I had tried to do. My uncle was on a secret mission. I did not want the word to spread and for the killer of my father to know that he was going to see the same fate as my father. I told her, “Can’t you see, mother? I went hunting.”

She said, “Son, it is too dangerous for you to be out by yourself. Your father has just lost his life. There are too many Iraqi troops around. You can’t be having any weapons on you. If they see you, they will shoot at you thinking you are a Peshmerga."

My mother was cooking in the front yard when I saw Beri at a distance. I thought she was going to get water from the well because she had a container with her. Instead, she came to our house. She greeted my mother and asked permission to talk to me. I could hear her, and I could feel my mother looking at me. I did not raise my head. I acted like I didn't hear anything.

My mother said, "Well Shivan?"

I did not answer.

She raised her voice again. "Shivan, you have a visitor."

I walked into the house without responding. I was very upset. I did not want to speak to Beri. "*After everything,*" I said to myself, "*this is how she treats me? No respect, no past memories.*"

It was time for lunch and my mother called us to eat. We all sat on the porch and ate. My mother looked at me. She knew I was upset. She did not want to start a conversation. She knew that I was short-tempered. I didn't want to start a conversation either. I started to talk to my siblings to change the strange vibe of energy flowing around. I did want to talk to Beri when she came. I wanted to ask her, why? I was not brave enough though to ask her what was happening. I said in my head, "*I will just try to find another girl.*"

CHAPTER 6

Changing Parties

The next morning, I saw Peshmerga from the Socialist Party all over my village. They were everywhere. There were about thirty of them. Some of them had made dug outs in the outskirts of the village, some were socializing with the villagers.

I did not bother.

Then, that after noon, a few of their Peshmerga came to my house. They asked my mother where I was. I was inside, but my mother told them I was not home. My mother came inside after they left. She said, "Son, please don't leave us again. We need you to help around the house."

As bad as I wanted to stay and support my family in the farm fields, I had to make some money. Farming was good, but it was hard work. It also didn't pay well. I told my mother, "Don't worry. I will see what they want."

I went to where most of them were based at the mosque. When they saw me, a few of them recognized me. They came, shook my hand and kissed me on the cheeks. Some of them jokingly said, "We found you. You can't run from us." I smiled and sat down with them.

My old commander, Hishyar, walked in, and I got up to greet him. He said, "Shivan, how are you?"

I said I was okay.

Hishyar said, "We have come to get you. I think your vacation is over, right?"

I smiled and told Hishyar what had happened. I told him that my family needed me to stay.

Hishyar said, "Shivan, you are a good fighter and very smart. We need you in our unit. We will leave tomorrow, at dawn. If you come with us, we will pay you double."

I was getting fifty dinars a month, about five hundred United States dollars, at the time. One hundred dinars would be equivalent to a thousand dollars. That was a lot of money. I needed that money to help my family. The offer was too good to say no to. I told him, "Okay. I will be ready."

He smiled, shook my hand, and gave me a kiss on the cheek.

That evening, I was getting my things ready for the next day. My mother found out from other villagers that I was going to leave again the next day with the Socialist Party Peshmerga. She had sent a word to my Uncle Azad. He also joined the Peshmerga but was with the Kurdistan Democratic Party (KDP). She begged him to come and talk with me.

The next morning, I went to the well to grab water for my mother to be able to cook and do other chores. I looked up north of our village and saw a never-ending line of men coming to our village. They were at a distance, so I couldn't tell who they were at first.

The Socialist Peshmerga, shouted "Hazirbin" (get ready). All of their Peshmerga came to the mosque. Something alerted them to get into their positions for a fight. It was a long line of men coming to our village. The Socialist Peshmerga got their weapons ready. Hishyar told them, "Okay, men, we have company. If you know they are trouble, shoot and don't wait for my signal. Just shoot as many as you can." The other villagers started to panic. Everyone rushed into their homes.

I went inside my house and told my siblings not to get out. As the line of men got closer and closer, I saw that they were also Peshmerga. I had never seen so many. They surrounded the village and I could see them at every corner. About fifty of them, heavily armed. I knew that it was the Kurdistan Democratic Party (KDP) Peshmerga. They were the strongest and had the most Peshmerga in their units.

They passed my house as I was standing in the porch looking at them. They went straight for the mosque. After a few minutes, I could hear talking between the leaders of the two Peshmerga parties. Sometimes they yelled at each other, but then things became calm after a few minutes. I could not hear any voices, just the sound of footsteps coming and going. I did not want to get involved. I did not know what was happening. If both of the parties had started shooting at each other, it would have been a blood bath, even though I knew that the KDP units would have easily controlled the situation. They would have taken all the Socialist Peshmerga under their control because of their bigger weapons and more men.

I heard a few voices nearing my house. I walked into my house because I did not want to run into any trouble. I heard a knock on my door.

My mother went to answer. She looked surprised and happy at the same time. "Oh, Sedeeq come in." It was Sedeeq, my father's cousin who of course was referred to as "mam" (uncle). My uncle Sedeeq was a known warrior among the Peshmerga. He was a very tall man with dark, tanned skin, a muscular build, a mustache, and a scary look. We call our elders "uncle," especially if they are our blood and from the father's side. Sedeeq was a KDP Peshmerga commander. He was in charge of all the Peshmerga with him at our village.

I got up and went to shake his hand. I said "Welcome, Uncle Sedeeq. Please, sit down."

As he sat down, my mother asked if it was okay to go outside.

"Yes of course. There is nothing to worry about. The other party was just leaving."

I stayed silent and waited for Uncle Sedeeq to finish. He said, "Those are our brothers. We are fighting the same enemy. We would not fight each other."

My mother brought us tea. We started to drink it and Uncle Sedeeq took out his cigarettes, offered me one, and lit his. He looked at me and said, "So, you're ready to fight?"

I did not know what he meant. I replied back and said, "Chewa?" (what?)

He said, "Are you ready to join our unit?"

"I thought I was too young to join the KDP."

"Well, I'd rather you be with my unit then with those Socialist Peshmerga. Yes, we are fighting the same enemy, but you don't know who those Peshmerga are. They are from all parts of Kurdistan. They could take you with them anywhere outside of Kurdistan Region of Iraq. You could be fighting in Turkey, Iraq, Iran, or Syria. In our unit, we are all from the same area. Everyone knows someone else. You will be protected and not worry about anything. You will never be too far from home."

I said, "Yes. That sounds great, Uncle."

Then Sedeeq continued, "The money is not good;. you are mostly volunteering. You will have good weapons, though, and enough food so you don't go hungry."

My mother interrupted and said, "Son, if you really want to be a Peshmerga, please go with your Uncle Sedeeq's unit. They are all people that know your father and your uncles will be with you."

I couldn't say no to my uncle, nor my mother. I answered right away. "Yes, Uncle. I will join the KDP Peshmerga."

Uncle Sedeeq, said, "Good. Here is twenty dinars. I will go get your rifle, a pistol, grenades, and ammo. The good thing about all this is that your task is to stay at the village and guard it from the enemy."

I was shocked. This was a great deal. "Okay mam."

He got up, called one of the Peshmergas, and told him to bring a rifle, pistol, ammo and some grenades. The Peshmerga came right away with everything he ordered. He handed them to me and said, "I have to get going. I will leave ten other Peshmerga here. When we need you guys, we will call for you. In the meantime, take care of the village, your family, and the Peshmerga I leave behind. Your father was not only a hero but a man who showed great hospitality."

"Yes, Uncle. I promise I will do that.

Before the KDP Peshmerga units left, I saw the Socialist Party Peshmerga forming a line to leave as well. As they were heading out the village, I ran to them. I caught up to Hishyar who was in front leading the troop.

He did not stop and kept on walking, but at a pace where I could talk to him. He said, "I should be more careful from now on. I can't just accept anyone. You have a strong family tie; you did not tell me that."

"I am sorry. I did not know that my mother had called for my Uncle Sedeeq."

Hishyar said, "I am just kidding. I understand, Shivan. This is your area; it is wiser to fight alongside your family than with someone like me."

"Please don't say that. You are my family too. You are my brother and so are all of these Peshmerga with you. They are all my brothers and sisters." I repeated the same but louder so all the other Peshmerga with him could hear.

Hishyar said, "Shivan you are an intelligent young man. Take care of yourself, and maybe I will see you again on the battlefield fighting the enemy."

I nodded. "Yes. Yes, we will see each other again. We won't stop until our land is free from Saddam's dirty Ba'ath regime."

Hishyar looked at me, smiled, and slowly disappeared in the line as other Peshmerga in his unit passed along.

I turned around and headed back to where the other ten Peshmerga were waiting for me down the hill. That was the last time I saw Hishyar. He was a good man, and I was going to miss him and the other Peshmerga whom I befriended with the Socialist Party.

I had to make the best choice and that was to fight alongside my family. This would be beneficial for me, my family, and my village. I was a KDP Peshmerga and this was something great. People looked up to you.

As I came back to the village with the

other Peshmerga, other villagers looked at me differently. They saw a leader. Little children came up to me and walked with me. They asked if they could be a Peshmerga. I smiled and told them yes, in the future. Young girls in the village were staring at me and smiling.

We walked by Beri's house. She was outside. She waved at me and smiled. I didn't wave back, just gave her a small smile. I made myself act like the man of the village. I was happy that I changed my party. It was a wise decision, thanks to my mother.

CHAPTER 7

Guerrilla Attacks

It had been months of no activity. I had not heard anything from my Uncle Sedeeq. The ten Peshmerga left with me settled in the village. They made a temporary house to sleep in. I stayed with my family at night, though I did provide the Peshmerga with everything they needed. We patroled the village together from dawn to dusk. We worked in the fields from time to time, helping other villagers with their farmland, and they shared their crops with us. We had not gotten paid for over three months. We were hesitant to do target practice because we didn't want to run out of ammunition. We needed to be ready just in case the enemy arrived.

One evening as we got settled in, I suggested to the rest of the Peshmerga that we should send out a scout to see what was happening to bring us current news, ammo, and our pay. They all agreed it was a good idea.

Zana, one of the Peshmerga who was about seven years older than me, said he would go. Zana was a small, young, good-looking guy that would be able to move quickly. Another raised his hand: that was Awar, and the last who wanted to go was Lawin. They were also small in size and could reach the Peshmerga at a fast pace. All three of the Peshmerga who volunteered to go were young, in their early twenties. I was worried that something would happen to them. They could run into trouble with the older Peshmerga if they asked for our pay because most of the older ones were volunteers. Other villagers might not take them seriously and not respect them because of their age. Lastly, there were only three of them, and they didn't have much experience in fighting.

I hoped for the best and prayed to God

that they didn't run into any Iraqi ambush. I asked the villagers to each contribute something for them to take along with them. Food, clothing, hygiene supplies, and blankets are what I asked of the villagers. A few of the villagers brought some supplies to them that I had asked of them. We all said our goodbyes and prayed to God for safety and protection.

They headed for my Uncle Sedeeq's bunkers, wherever they were in Berwari Bala. I wrote a note to my uncle, asking him to help us with supplies and asking when we might be needed for an assignment. There were eight of us Peshmerga left in the village.

The next day, midafternoon, I heard helicopters coming. Everyone in the village panicked. Villagers in the fields and mothers with their crying children ran out of their houses, others running to shelter in the caves uphill from the village.

I saw my mother in front of the house with my siblings crying. I hurried to them. I escorted my family to the caves up the mountain, and the other Peshmerga helped the other families as well. The helicopters were getting closer and closer.

Then I heard gun shots. I could see three helicopters coming towards our village.

Minutes later, they fired at the village. They were shooting at our houses, animals, and anything else that moved. These raids came often, but most of the time it was one plane or helicopter. This time, they must have gotten some information about a Peshmerga presence in the village. They fired on the village for about ten minutes then flew south to the town of Bamerne near Duhok City. That is where the big Iraqi military air base was located in the Duhok area.

Even though the helicopters were gone, everyone was still frightened to come out of the caves. The children were crying. It was not a pretty situation. I couldn't do anything but just stare at the village. While things were up in smoke and some things on fire, we waited until dawn to go back to the village. Planes and helicopters usually came during the day.

When we got to the village, there were bullet holes everywhere. A lot of the farm animals such as sheep, chickens, and cows, had been slaughtered by the bullets. We ate their meat the next day at the mosque. All the villagers came. Each family had cooked something. We were all happy to be alive and that no one was hurt. We talked about what might have happened. Did the Iraqi army know that the Peshmerga was stationed in Belizan? Did someone give them information about us? Were Zana, Awar and Lawin okay? Did they get arrested and tell the Iraqi army where we were? People brought up different things that might have ticked off the Iraqi army. I just stayed quiet.

Finally I said, "Alright everyone, that is enough. We just have to wait until Zana, Awar, and Lawin return. We can't make assumptions. We must wait to see what message they bring us."

Late at night, as we were about to fall asleep in the village, we heard footsteps coming. It sounded like a lot of men. I slowly shut off the kerosene lamp and grabbed my rifle. I stood beside the door and pointed my rifle to it. Whoever opened it would be shot right there. I heard a voice calling. I did not hear clearly at first what was said, then I heard it again.

"It is us, the Peshmerga." They

repeated themselves again and again, "It is us, the Peshmerga."

Most of the children were sleeping, but the adults left their houses as did I. It was my Uncle Sedeeq's unit. He brought his large force once again. There were over three hundred Peshmerga with them. Zana, Awar, and Lawin were also with them. My Uncle Sedeeq came to my place to stay. The other Peshmerga went to the other villagers' houses. Most of the Peshmerga had camped outside to keep alert for the Iraqi army.

Uncle Sedeeq and I had some long conversations. He told me what happened, where they had gone, and what was happening. I heard a knock on the door, and it was my other two uncles, my father's brothers, Azad and Checkdar. There were other Peshmerga with them because they were staying with them. My mother greeted them and went out to make tea for us. We sat down and talked until we could see the first light of day.

Uncle Sedeeq handed me money, saying "Sorry it's late."

It was over sixty dinars. It wasn't much, but being from a village, it was a lot. I was happy and handed it to my mother for safeguarding.

Sedeeq said, "Tomorrow, we have to leave early, all of us Peshmerga here in the village. We are going to do some major guerilla attacks. The time has come.

"The Iraqi army has been notified that there are over twenty thousand Peshmerga armed and ready to fight. They are attacking every village in the Kurdistan region. They are arresting people in the city left and right to get any information on the movement of the Peshmerga. They are harassing people and accusing them of being Peshmerga when a lot of them aren't." The Kurds were being sent to jail and executed. A lot of families were asked where their men were in the household. If the family's men were not home, they were automatically arrested and relocated to Arab cities in the south. My Uncle Sedeeq said, "We must attack all of the Iraqi bases starting tomorrow. We have got orders from the KDP headquarters. The Iraqi government is blaming the Kurds for aiding Iran as well."

At this time, the Iraqi government was not only fighting the Kurds but Iran as well. The Iraqi government was also killing their own kind, Arabs, but ones who were a different religious sect from the Sunnis, the Shi'ites in the south of Iraq. Saddam was killing many innocent people, and it was time to get our revenge.

We slept through the afternoon. Many of the Peshmerga knew that they were going into battle. They enjoyed themselves in the hot summer day. Some of them were bathing, some were barbequing, and some were singing songs and dancing.

I was sitting in my front porch with my Uncle Sedeeq and some of the other Peshmerga. I saw Beri by the water station. I had to go and say hi to her. It had been a long time since we talked. I got close to her. Before I would leave, I had to say goodbye just in case something happened to me. I wanted her to forgive me, forgive me for ignoring her for so long. I was not mad anymore. I was happy. I was happy for her.

I excused myself and went to the water station. There she was, looking beautiful as always. I asked if she needed any help.

She looked up, surprised, and said,

"No thanks." Then she turned back to her chores.

"Beri..."

She did not reply.

"Beri."

She looked up at me with her beautiful eyes.

I took a deep breath. "I am sorry. Sorry for ignoring you."

"You should be."

"Why should I be sorry?"

"You should be!" she said again.

I got frustrated for a second, but I held my temper and raised my eyebrows in question.

She said, "I am not getting married to that idiot. I don't even know who he is. My father said okay to them before he asked me. I don't know what is going on, Shivan, I tried telling you that, but you didn't listen. You didn't give me the chance to explain myself."

I told her I was going to go kick her father's ass. I started to walk off but she said, "Stop, just stop. It's okay. My father has agreed with me. He told me that his wishes were for me to marry that guy. That his family was wealthy, and they would help ours, but he told me, I am sorry. You can say, 'Yes,' when you are ready."

My temper had come down. I felt relief and like something heavy was off my shoulders after hearing that from Beri. I walked back to her."I am leaving to fight with the Peshmerga. When I return, I will ask for your hand. What will you say, Beri?"

She looked at me with delight. "Promise?"

"Yes, I promise. Heshtetkim, Beri" (I love you. Beri).

She replied in a low voice, "Heshtetkim, Shivan."

I said goodbye and went back home.

That evening, I prepared myself for battle. I told my mother to pack me some clothes and put them into my backpack. She did that along with some things to sleep on and to cover myself with. My mother packed dried food as well for me to take. I cleaned my rifle and pistol and made sure all my gear was ready for the next morning when we would leave the village.

I spent time with my mother and told her what needed to be done and discussed who I was going to hire to take care of our farmland while I was gone. My mother didn't need help. She was already a strong lady. My siblings were young, though. She needed to be around them most of the time. They needed her and she needed to be there to watch over them. I told her not to worry, that she could focus on the house chores, and I would pay another villager to look after our farmland.

She agreed but had tears in her eyes. She was scared that I might leave and not return. She had already lost my father and didn't want to lose me. She said in a quiet voice, "Son, please be careful out there. Don't try to be a hero. Just do as you're told and try to stay behind the Peshmerga. They already have experienced fighters. You don't have to prove anything to anyone."

I told her, "Yes mother, I will do that. Don't worry about me. I will be okay."

She excused herself as my younger siblings were crying. She went to get them ready for bed.

I laid in the living room by myself. I started to think about my future. What did I want to do? Where did I want to go? How I was going to make money after the war and provide a nice home for my family, either in Zakho or Duhok?

Life in the village was tough. The city was where the money was at, where the jobs were. There was always something to do. Don't get me wrong, I loved my village. Being in the mountains, in the farmland, a person was so free. It was lonely though. We didn't have all of the luxuries that city folks had. They had electricity, TVs, cafés, restaurants, and endless entertainment. I wanted to give that to my family.

It was early in the morning, barely any light, when I heard a pounding on my door. I went to it and called out, "Yes?"

A familiar voice said, "It's me, nephew, your Uncle Sedeeq." I opened the door. He looked sharp, ready for war. He had his Kurdish khaks, or tradtional Kurdish clothing, on. His rifle was on his shoulder. His magazines, grenades, and pistol were wrapped around his waist. His Kurdish kulav and dersuk turban were tightly wrapped around his head. He said, "Are you ready?"

I had been ready. I barely had any sleep that night. I was thinking all night about my future, Beri, my mother, my siblings. I said, "Bele mam" (yes, uncle).

He stepped out for a cigarette as I grabbed my rifle and other supplies that my mother had packed. Then I saw my mother. I could not help it. When I saw her precious innocent face, I started to bawl. I cried and went to give her a hug. She cried too. She walked me to the front porch and greeted my Uncle Sedeeq. She said, "Xode gel webit" (God be with you).

We started to head out of the village, a three-hundred-strong Peshmerga unit. As early as it was, all the villagers were out. They were there to support us and to say goodbye. Even children were up early and came up to us and wanted to shake our hands. As I walked past Beri, she looked at me, smiled, and gave me a secret kiss with her two fingers.

Going into battle was a scary thought, but the support from my people gave me hope that there was a better tomorrow, that this fight was worth it. This was everyone's fight to be free. We, the Peshmerga, were defenders of our homeland. We wanted more rights for our people. We wanted to be free just like other nations. We were fighting for the Kurds and Kurdistan.

I had my head held high. I knew that I was meant for this. I knew that my father would be proud of me, despite him encouraging me to get an education instead of being a Peshmerga. We waved one last time at the villagers before disappearing into the Metin Mountain.

CHAPTER 8

Getting Married

I was finally going back home after a long time away in the mountains. I had seen it all: continuous fighting, death, being afraid, insomnia, hunger, being cold, and getting homesick. I was ready to see my mother and siblings. During these years, I missed Beri so much. I would wonder what she was doing and how she was doing? There were many pretty girls in the villages we visited, though I looked at all of them as if they were my sisters. We were their protectors, and I could not look at them in a wrong way.

I just wanted to rest and get as much sleep as I could. I hadn't shaved for about a year. My khaks were ripped up, my boots had holes in them, and I barely had any ammunition left for my rifle. I needed to rest and have some time off. I told my commander, who happened to be my Uncle Sedeeq at the time that I needed to go home. It was almost a year that I had been gone. I needed to go and see my family. He agreed and said, "Go."

Coming back home was the tough part. I was not with the Peshmerga unit anymore. I was all by myself. I could be killed any second by a plane, helicopter, an Iraqi army ambush, a mine, or wild animals. It took days for me to arrive back home.

The two days felt like the whole year I was gone. I arrived at the east side of Kani Hishke (the Dry Well). This was where my relationship with Beri had begun, years earlier. There, I put my backpack down and took my clothing off slowly. It felt like it had been glued to my body. I wanted to clean up before I went to the village.

The villagers all knew each other. They knew how things worked. People always shared supplies because that was just the traditional custom and our way of life. We did this by leaving things where others might need them in the future if they knew where to look. There, at Kani Hishke, I was able to find soap, a mirror, and shaving supplies. There was a prayer rug, too. I never really prayed much, but after I got done cleaning myself up, I prayed and thanked God for my safe return home. I then headed to my house.

I finally felt refreshed, better. I was really excited to hear the sounds of children, women, and men coming from the village, doing different tasks. As I was walking into the village, the local villagers I ran into came up to me. They shook my hand, kissed me on the cheeks, and gave me hugs.

The kids would come up to me and ask me questions: how many Iraqi soldiers I had killed? Had I seen any dead people? Could they join the Peshmerga? I smiled at them without answering and kept on walking. Other young kids went running to my house to give my mother and siblings the good news that I was back home.

As I came near my house, almost the whole village had come to my house to greet me. As I walked to them, they all had beautiful smiles on their faces. It was the most memorable moment of my life. I felt like a hero, as if I had somehow saved them. One by one, the villagers came and hugged me, kissed me on the cheeks, shook my hand, and thanked God for my safe return.

My mother started bawling when she saw me. She and my siblings came running to me and held me tight for about five minutes. My mother kissed me on the hands, shoulders, eyes, and forehead. She was very happy to see me.

Beri's parents came up to me as well. Her dad shook my hand firmly, and her mom kissed me on both of my cheeks. Beri came up to me and said in a shy low tone, "Alhemdullilah tup silameti geheshti" (thank God for your safe return).

All the villagers were trying to usher me into their houses. They invited me to eat supper with them and drink a tea. I told all of them no, that I was tired and would see them later. I needed to rest. I sat down with my family for a few hours before going to bed.

My siblings were all around me. They had missed me a lot and I had missed them too. Not having my dad around was tough. My siblings needed a father and I had to play that role, even though I was young. I did not act my age; I acted like my uncles, the way they talked, the way they dressed, the way they went about their daily activities. People looked up to me now that my father was gone. When they saw me, they would tell me, "We see your father."

Now that I was back home, I needed to keep my promise to Beri. I needed to ask for her hand. How I was going to do it was the tough part. I could not go to her father myself, of course. I had to send an older man, someone respectable, someone to represent me and my family. Her father didn't really like my uncles, Azad, Rojhat, and Checkdar. He knew them as troublemakers. If I would send them to ask for Beri's hand, her father would say, "No," instantly. I had to come up with another plan. Perhaps if Azad, Rojhat and Checkdar went, it would be okay as long as there were some respected elders with them that had well-known names to tag along.

My mother was working too hard. We needed some help around the house. Like I mentioned, my siblings were too young and mother's chores were too much. I had a plan. I would write a letter to my Uncle Sedeeq, my Peshmerga commander. I would tell him that I wanted to get married and was asking for him to help me. He could get the elders in Brifka, my original village, to come to Belizan. He could convince Beri's father that I was a good guy, to give her to me because I was from a well-known family, and that I would take good care of her. I pulled out a piece of paper and started to write.

> Dear Uncle Sedeeq,
>
> I want to thank you for sending me home. I am now with my family. They are all doing well. Things are starting to go well again. I am back working in the fields. I am helping my mother around the house. There is one thing that I want to ask you, though. I am in love with a girl in my village. Her name is Beri, Dijwar's daughter who moved to Belizan. I know you know him very well, and he knows you. I am asking, in the name of our family, if you will help me ask for Beri's

> hand. I know that Azad, Rojhat, and Checkdar are all here in the village, that they are my first uncles, but to be honest with you, I feel like you have more power to make this happen, along with our other relatives in Brifka.
>
> I am waiting on your response. Thank you, and God be with you.

I sent the letter with a few Peshmerga that were stationed in Baze village, about a twenty-minute walk from Belizan. The Peshmerga were not my mailmen, but I asked kindly if anyone was going to Amediya, which is where our KDP headquarters was. I was sure that if the letter was sent there my Uncle Sedeeq would receive it. A couple of the Peshmerga said, "We are going to get supplies there. We can take your letter."

I was very happy. I went to the local market in Baze and bought them cigarettes and drinks. I also gave them a few dinars for taking the letter. They refused to accept my money, but I put it into their pockets. That is how we did things. You treated everyone like they were someone. This gave a person value and worth.

I came back to Belizan late that afternoon. I stopped by the farm field and looked at our crops. Everything looked good. Things were starting to grow. We had a garden full of tomatoes, cucumbers, Kurdish parsley, cabbage, lettuce, cantaloupe, watermelon, onions, and many other fruits and vegetables, not to mention the rice and wheat that were fully grown. Our orchard trees were in good shape, and we were able to harvest apples, pomegranates, grapes, figs, walnuts, peaches, and almost every other fruit you could desire.

I went home about an hour later. My siblings were taking their afternoon nap. I told my mother that I needed to talk to her. She came and sat next to me in the porch. She asked if I wanted something to drink. I said, "No."

"What is it son? What is on your mind?"

"I am going to ask for Beri's hand. I want to marry her."

She looked down. "Who is going to help you, son? Beri's father doesn't like our family, and we don't have any money."

"That is not true, mother. Beri's father

loved my father. They were friends. He doesn't like my uncles, but that doesn't mean he doesn't like us. They are the troublemakers, not us. We have some money, mother. We don't have to have a big wedding."

"So how do you plan on marrying her with your uncles doing the asking?"

I told her that I had sent a letter to my Uncle Sedeeq asking him to come to Belizan to talk to Dijwar and ask him for me to have Beri's hand in marriage. He would bring along Azad, Rojhat, and Checkdar, but Sedeeq would do the talking. I also told her that we would bring other relatives from Brifka village, where most of them were."

"Okay son. I am just worried that her father will still say, 'No.' I don't want you to be sad, son."

"I won't be sad if he says, 'No,' but Inshallah (God willing), he will say, 'Yes.'"

My mother thought for a moment, then asked, "What about your Grandpa Reshid? Maybe he can help."

"Yes, that is a good idea, but that grandpa is in Mosul, working. We will send him a letter telling him what we are going to do. Maybe he can come for the wedding day if everything goes as planned.

That evening I started to write to Grandpa Reshid, my dad's father.

Dear Grandpa,

I hope that you are doing well. We miss you a lot. I am back home for a while. I am writing because I am planning on getting married. I want to marry Beri, Dijwar's daughter, from Belizan. I would appreciate you helping me to ask for her hand, but I know that you are working in Mosul and won't be able to. Hopefully if everything goes well, you can come for the wedding date. I will ask your sons Azad, Rojhat, and Checkdar to go with me. I have also written a letter to our cousin, Sedeeq, to ask other elders from our village in Brifka to help me. I hope that you approve of this, and we hear from you soon.

My mother says, "Hello," and

that she hopes that you and my grandma Fatima are doing well.

God be with you.
Your grandson,
Shivan

It had been over two weeks that I had not heard from my Uncle Sedeeq. I told my uncles Azad, Rojhat and Checkdar about my intentions. They were happy about me wanting to get married but weren't too happy with my decision choosing Beri. They did not like her father, Dijwar, either. They were in constant arguments in the village with him. I did not want to hurt their feelings and didn't want to tell them the truth. I just told them that it would be a good idea if all of our relatives were present. This would make us look good, and Beri's father would be more convinced to say, "Yes."

Azad was always coming and going. He was a Peshmerga as well, a well-known fighter that was called on frequent missions. He did not go for months into combat, though. When there was damage that needed to be done, he was called into action. Therefore, I could not rely on him when the time came. Besides he had serious anger issues. If Beri's father would say something out of the ordinary, he would probably pick a fight with him right there.

My Uncle Rojhat was also a man who got angry easily. He held grudges against almost everyone he came across. For some reason, in his own little world, he was trying to take after my father, being looked up to and respected. That wasn't going very well for him.

I knew that Beri's father, Dijwar, couldn't stand him and did not want to be in his presence. Dijwar knew that my Uncle Rojhat was fake. Therefore, I could not rely on him either. Rojhat didn't show you that he was mad at you, he would just not talk to you and ignore you. Not just for days or months, but for years. I know that he hadn't talked to Dijwar since he was in Belizan.

My other Uncle Checkdar was the type of person who just went with the flow, whether it was good or bad. He just agreed to everything. I was afraid if he came solely that if Beri's father said, "No," he would just say, "Okay, thank you for your time," and leave.

I wanted a negotiator. A type of person who would be able to explain why I am the best person for Beri. Why it would be important for both of our families to come together. Don't get me wrong, my uncles are great people in a group setting, but not by themselves. This is why I thought Sedeeq would be the best option as the lead negotiator in convincing Beri's father to let her marry me.

One late afternoon, I heard a loud noise coming from the farm field. There were men screaming and shouting at each other. I stepped out and all I heard was swearing. I knew that there was a fight going on. Who was fighting? That was the question.

My mother came out and said, "Shivan where are you going?"

"I will be right back."

"Please don't go."

"Mother, I will be fine. I am just going to see what is going on."

"Okay, but please, son. Don't do anything foolish."

"Mother I am not even involved. I am just going to see what the big fuss is all about."

As I started to walk toward the field,

I saw children running to the village. I stopped a few of them and asked what was going on. A few of the children said, "Your uncles are going to get a good beating." Then they ran off. The children belonged to the other villagers in Belizan. After hearing this, I picked up my pace and started running fast to the farm fields.

When I arrived, I saw that there were over twenty villagers from Belizan fighting my uncles, Azad, Rojhat, and Checkdar. I went to them and tried to break up the fight. I was hit in the back of the head by a few punches. I did not want this to happen, but I turned around and started swinging back. I grabbed a big rock from the field and tried swinging it at the other villagers. One of the villagers smacked the rock out of my hands with a shovel. We fought our battle but got a little few kicks and punches all over. There was just too many of the villagers.

Sooner than later, almost the whole village was there. The women, men, and children were all swearing at us. Luckily, there were a few good villagers that broke us apart. They grabbed us and took us back to the village. The villagers kept on trying to fight us as we were leaving. I think they wanted to kill my uncles and because I was their blood, they took their anger out on me as well. They thought I had come to help my uncles when I had no idea what the fight was all about in the first place. This was our story of Belizan village, always drama with the villagers, always fighting one another over whose turn it was to water the fields, whose shovel was missing, fighting over land. The drama between us and the local villagers seemed like a normal thing because it was so frequent.

On the good side, my grandma was remarried to my grandpa. She had one child from her prior marriage before her husband passed away. Her child's name was Mehmid. Mehmid found out what had happened to us, and along with his sons, he had gone to the fields to confront the other villagers, but they were no match for the other villagers because there were so many of them. They got a little beating too. My grandma's sister was in the village married to another villager from Belizan. She had walked all the way to Brifka to tell the rest of our family to come to the rescue, because the men in Belizan were trying to kill the Brifkis.

That evening, my extended uncles and dad's cousins came to Belizan. There were about thirty men from Brifka. They walked around the village, kicking the villager's doors, and swearing out loud to the villagers to come out and fight. No one dared to come out. My family members from Brifka shouted and shouted, and it seemed like Belizan village turned into a ghost town. The villagers thought they would get away with beating us. The men from Brifka had managed to catch a few of the villagers that were on the farm field and gave them a good beating to remember.

They came to my house after a few hours, showing their presence. I later found out that the whole fight was about water. Who would get the water first to water their fields? My uncles Azad, Rojhat, and Checkdar said it was their turn, and some of the other villagers from Belizan said that it was their turn. That is how it all escalated into the fight. There had been many other fights prior to this, but this was the worst of them all.

My mother made tea for them. It was

time for supper, and she told me to get some supplies ready. Rojhat told my relatives from Brifka that they should grab the people who started the fight and give them a good beating. Checkdar stayed quiet the whole time. Azad, who was the oldest now, said, "No, we are not going to do anything. The fight is over. That is it."

My relatives insisted that we leave the village before something else happened. Nobody wanted to say, "Yes," but it seemed to be the best option. The fight would only escalate to something worse. God forbid if someone got killed. My relatives crashed for the night. They slept wherever they could until morning came.

The next morning, we packed and were heading to Girka village, which was about thirty minutes of walking from Belizan. My dad's cousins, Selman and Hakeem, walked around our village, Belizan. They were swearing at the villagers to come and fight, but no one dared to come out of their home.

Our relatives guided us to Girka village. It was a sad situation to leave, but the wise elders knew that if we stayed, things would get worse. I did not even get a chance to tell Beri, "Goodbye." Her father was not home. I wonder what his reaction would have been if he was there during this incident. I am glad, though, he had not been home.

We spent about a month in Girka village. It was tough living there because between my uncles and my family, we shared only two houses which each had three rooms. Luckily, my siblings and cousins were all young. Word had spread throughout the Berwari Bala area about our incident. Many had our backs, and a few supported the villagers back in Belizan. All my family from Brifka knew and came to visit us from Brifka, Zakho, and Duhok.

It was late in August, and fall was coming. That meant we had to harvest our crops and orchard trees in Belizan. Was it the right time to go back, though? My uncles talked about it and Rojhat really wanted to go back.

We never understood him. Rojhat wasn't much of a worker. Why did he want to go back? I know that he missed home and wanted to work on his farm, but the intention was to start drama once again, for which he was well known.

Uncle Azad was the impatient and grumpy one. He had serious anger issues but for some reason he thought things through. My other Uncle Checkdar just said, "It's up to you guys if you want to go back. Whatever you say, we will do that together."

Azad advised us that we should call on our cousins from Brifka to come and visit us in Girka. Here, we would talk about going back, so they would be notified. He suggested it was best that they know, and that we get their advice before going back to Belizan. What if something else happens? There is only four of us and many more of them. We need our relatives in Brifka if something bad were to happen again. I agreed. Rojhat was upset and didn't want to do this, but he had no other choice because Checkdar agreed as well, which was a three to one vote.

Azad told us that he would head to Brifka first thing in the morning, a fifteen-minute walk from Girka village. We told him that we would go with him. There were many Iraqi troop ambushes and checkpoints. It was a dangerous time and we worried that it was best to go together. He laughed and said, "You don't have as much training as I do. Remember, I am Azad. The Peshmerga know that name." We laughed together and wished for the best.

The next morning, I woke up and Azad was gone already. I hung around the house and did some chores for my mother. Later in the afternoon, I went to visit my uncles Rojhat and Checkdar, who were living together. Azad's family and ours were staying together. They were a few houses away.

Later in the day I heard a lot of men's voices. We got up and saw that it was our elder men from Brifka. They greeted us and shook our hands with kisses and hugs. My aunts and mother hurried to get supper ready.

The elders sat down with us. They asked us about our living situation and how the rest of the family was doing. We had a lot of complaints.

My mother called to me, "Come here son. The tea is ready."

I went to get the tea from my mother and passed it around. There were about ten men from Brifka, so I had to make more than one trip back and forth.

While drinking our tea my Uncle Azad

said, "Dear cousins and uncles. We are happy that you are here with us today. The reason we have called you to come is that we want to go back to Belizan. Our homes are there, our crops need to be harvested, and there is not enough room to share houses in Girka. We need to go back to our village."

The elders suggested that it was too early. That we should wait for a few more months.

Rojhat got impatient. He got up and said, "Two more months? Nothing happened. It was just a small argument and a couple of fists thrown at one another."

Checkdar said, "Yes, I want to go back too. I can't take care of my family here."

The wise men were in silence when my Uncle Tayib asked, "How about you, Shivan? What do you want to do?"

I said, "I thank you all for coming here today. I know that you really care about us because we are one blood. However, we have invited you because we don't want to do things on our own. We don't want you to come around tomorrow saying that we did not ask for your advice or notify our family in Brifka that we were going back. I believe it is best that we go back. It has been over a month that we have been in Girka. My grandpa is supposed to be back in mid-September. Maybe he can calm the situation too. He married my grandma, who is from Belizan. Maybe they will listen to my grandpa and if not, they will not do anything else because of my grandma. We thank you again for coming to visit us. We will go back this Friday and will arrive in time for the Friday prayer in Belizan." It was Wednesday. We had two days to get ready and pack to head back.

Our relatives from Brifka ate dinner. They smoked their cigarettes and drank their tea. They said, "You guys can go back under one condition: that whatever happened is in the past. Moving forward, everyone must get along. We don't expect you to be friends but just respect each other as being in the same village and being neighbors."

We all agreed. Rojhat said in a funny, whispering voice, "Those motherfuckers."

Everyone burst out laughing. A couple of the elders said they would come along on Friday, to bring us back to Belizan. The rest left before nightfall.

During the two days prior to heading back to Belizan, we thanked the villagers from Girka for hosting us, for treating us with respect, and showing us great friendship. Their hospitality through our tough time was very kind. After all, we were from the same big tribe in Berwari Bala. In a way, we were all related, and we were there for each other when a village was in distress in Berwari Bala area.

Friday morning came. We got our donkeys packed with our house supplies. It was early, too early, but we didn't want to miss the Friday prayer. We headed out for Belizan.

The twenty-minute walk lasted over an hour. We had children with us, and the donkeys are not like horses. They don't mind rough terrain, but they are very slow. We had a good time though, talking all the way, taking rests by fresh streams of water, and enjoying some Kurdish snacks such as kades (dates and coconuts wrapped with dough) that we brought along the way.

We reached the peak of Belizan in Hetrishe. Hetrishe was a beautiful area, full of tree shade and water. The mountains and hills of Hetrishe could be seen from a distance. There, we rested again, washed up, and the elder men prayed. So did my mother and aunts.

We came into Belizan. Villagers were standing in their porches. Some of them were just waking up and some having breakfast. They looked at us as we walked by. They did not say anything, and neither did we. We came by the mosque and each of us headed in different directions to our houses. We had to unpack, eat something, and go to the mosque to do Friday prayer when it was time. I helped to get the fire started for my mother to make us food. Then I went to the well to get some water for her.

When the Mullah Imam Sheban called to prayer, I washed up and headed for the mosque. Sheban was a kind soul. He was very religious and friendly. I wasn't really a religious guy and didn't go to the mosque every Friday, but this was important. We needed to show our presence, to let the villagers of Belizan know that it was our home, too. We had land there. Most of us were born there. It was our village too, despite us being from Brifka.

I walked into the mosque and saw

that my uncles were present, along with some of the men from Brifka. We did the Friday prayer, and as we were walking out, the villagers from Belizan came to shake our hands. We started conversations in front of the mosque as if nothing had ever happened. I had butterflies in my stomach, and yet felt good that we had made peace.

I went home and told my mother. She became very excited. She feared all the time because of losing my father. She did not want anything bad to happen to our family again. What was especially great that day was that I saw Beri's father, Dijwar. He was also at the Friday prayer. This was a good opportunity for my uncles in Belizan and those from Brifka, who were in Belizan, to help me.

My family from Brifka stayed with us for about a week. A couple were at my uncles and one of them was staying at my house. That afternoon we heard kids hollering "Peshmerga, Peshmerga hatin" (the Peshmerga are coming).

Which Peshmerga? There were about fifty men, and as they got closer and closer, I knew that it was my Uncle Sadeeq.

I was full of joy and rushed to my uncles houses to tell them.

As usual, the first thing the Peshmerga did was go to the mosque. It had the most seating space. It was also a gathering place for the villagers to meet them there, and then to host them at their homes if they chose to. All of us villagers in Belizan had gathered at the mosque to greet the Peshmerga. The Peshmerga came with my Uncle Sedeeq leading them in the front. He came shaking the hands of the villagers one by one.

As he got closer and closer, he looked at my uncles, looked at me, and smiled. He approached my uncles first and shook their hands. When he came up to me, he did not shake my hand but rather grabbed me and gave me a big hug. He kissed me on both of my cheeks.

"Bi Xer hati mam" (welcome, Uncle), I said with a smile.

He said, "Sorry for the delay, nephew. I received your letter two weeks ago and tried my best to come as soon as possible. You know the situation."

I said "I know, Uncle. I am glad you came. My mother will be happy you are here."

The villagers each escorted the Peshmerga to their houses. My Uncle Sedeeq came to my house. That evening we sat down. He told stories of his battles and why it took him so long to come. He said, "Shivan, I had great respect for your father. He was my dear cousin. I still can't believe he is gone. I know one thing though. As long as I am alive, I will make sure that nothing happens to your family. I don't have much but whatever I have, I promise to share it with your family. I have a family too, but you are our orphans, and our religion says to protect orphans at all costs." He shifted in his seat and took a sip of tea. "Okay, so you want to marry Dijwar's daughter? What is her name?"

"Uncle Sedeeq, Her name is Berivan."

"What a great name. Shivan and Berivan. What two great names that are meant for each other. What about Azad, Rojhat, and Checkdar? Have you talked to them?"

"Yes Uncle, I have talked to them, but the person I think can make this happen is you."

He smiled and said, "Well, thank you Shivan. I am flattered to hear that."

"Uncle Sedeeq, let's be honest. You are a respected man. All these Peshmerga with you today shows that people will listen. You know that Dijwar doesn't get along with your cousins."

He said, "Yes, I know. Dijwar is a picky man. He doesn't like troublemakers. I mean, I don't think my cousins are that bad, but they do have a bad reputation in this village."

I said, "Exactly, Uncle. This is why I wrote to you. Please, you have to help me."

"Shivan, I will help you. This is why I am here today. What I want you to do is to go get your uncles Azad, Rojhat, and Checkdar. We will talk tonight about our plan to ask Dijwar for his daughter's hand. But before we do that let's ask your mother." Uncle Sedeeq beckoned her over to us. "Aysha, has Shivan talked to you about Dijwar's daughter, Berivan?"

She smiled and said, "That is all he talks about."

We all started to laugh.

I excused myself and stepped out to go get my uncles and my other relatives who were staying with them. I knocked on each of their doors and invited them for tea. They all agreed and came with me. We walked back to my house.

That night Sedeeq, Azad, Rojhat, Checkdar, and the five of our relatives from Brifka all talked about my situation. Sedeeq opened the conversation first.

"My dear cousins, tonight we are here to talk about our nephew, Shivan."

Rojhat said, "Xere Inshallah" (good deed, God willing)."

Sedeeq replied back to him, "He said there is good in this conversation. Our nephew, Shivan, wants to marry Dijwar's daughter, Berivan."

Azad said, "Xosh shole" (good work)."

Rojhat said, "Ma che kichen di ninen?" (aren't there any other girls?)

Checkdar said, "Keyfa wi ye" (it's up to him).

My Uncle Sedeeq said, "This is what we will do. We need to have a big feast tomorrow evening. We need to butcher five sheep. Shivan's mother Aysha will cook for us, along with your wives, Azad, Rojhat, and Checkdar. Your wives can ask some of the other women in the village for help. We won't tell anyone of our plans and what we will do until after we eat tomorrow. We will invite Dijwar over along with some of the elders in Belizan village."

We all agreed that night of our plans for the next day to come. I had butterflies in my stomach that night.

I woke up extra early the next day. I barely got any sleep. I was very excited as well as nervous. No one else was up. I prepared the fire and put the tea kettle on it. I grabbed my pack of cigarettes that my uncle Sedeeq had given me and smoked one until my tea was ready. *What would Beri think if I told her beforehand? Would she be super excited or nervous, or maybe both just like me?* I decided that day I would show myself to Beri, but I would not tell her what we were planning. I wanted this to be a surprise. She knew that I was going to ask for her hand someday, but that day was nowhere near in her mind.

Tomorrow the men were going to eat, drink tea, and smoke their cigarettes. After a few stories here and there, they would bring up my situation. Then they would ask the father for the bride's hand.

Usually, the groom is not present. He can be, but it's very unusual because it's kind of an awkward situation. The groom-to-be is already very nervous. Having to be present would bring more horror to the scene. "*I would eat with the men, serve them tea, and excuse myself*" I thought. Since most of the men were going to be at my house, I would find a way to sneak out with Beri. We would go somewhere and talk until we heard the good news.

Or the bad news.

We were all very busy the following day. Those of us in our family who knew what we were going to do brought the sheep for slaughter and the wood and other supplies from Baze village. There wasn't a big market in Belizan. Choosing where to shop was complicated. The nearest place to get most of the items was either in Baze, Bamerne or Begova village. Belizan and Baze had limited inventory, unlike Bamerne and Begova. However, Bamerne and Begova were dangerous because of the Iraqi troops' presence and large bases. Bamerne and Begova would have given us weeks of supplies, but we didn't have the time nor the capability of going in and out safely.

We didn't have much time and we needed our supplies fast.

My mother and aunts went to Baze to get supplies. The men butchered the sheep. Uncle Checkdar and I went to get wood. We needed to make everything perfect. Asking for someone's daughters' hand was not a joke. Even though many people didn't have much in terms of money, they found ways of finding supplies to make celebrations and special occasions happen.

Everyone was working very hard. My mother and aunts came back early that afternoon. They got their pots and pans out and started making food right away. Some of the men in Belizan asked my uncles what the big occasion was about. They would say something like, "Just a little get-together." The women came to ask my mother and aunts why they were so busy cooking food. They would reply, "We are making food for the Peshmerga."

It wasn't a lie. The Peshmerga were just not in Belizan itself. They were camped on the outskirts of the village. My Uncle Sedeeq's Peshmerga knew that he was in Belizan on business, and they did not want to intrude on family affairs. They, therefore, spent most of their days in the outskirts of the village, even though they came back during the night to sleep at the mosque or in other villagers' homes.

We could not tell anyone of our plans. We lived in a small village. Everyone wanted to know everything. There were many feuds between us and the rest of the villagers in Belizan. Even though it was my uncles' fault most of the time, they were still blood. I had to side with them because that is just how things went. We were afraid that if anyone knew of our plans outside of our family that they would talk Dijwar out of it. Most of the villagers were not employed, they were not Peshmerga, and they didn't even farm well. They were good at gossiping, though. Gossip leads to arguments, then to fights, and sometimes even death. Gossiping is not the way to a life of peace and good relationships, but I guess that is how some people wanted to live their lives.

It was nearing the evening. Checkdar and I brought plenty of wood. My mother and aunts cooked most of the food. The sheep had been butchered a long time before, and the ladies were cooking them already. My uncles from Belizan and Brifka all came to my house and sat on the front porch along with Uncle Sedeeq. They all sat down, drinking their tea while smoking their cigarettes.

Uncle Sedeeq spoke up, "Dear cousins, I believe the time has come. Azad, Rojhat, and Checkdar. Please go and invite some of the elders in Belizan. Checkdar, I want you to go to Dijwar's house and invite him. Don't tell him we are going to have supper. Just tell him Sedeeq is here and wants to talk to him."

As the sun was about to set, the elders in Belizan came to my house and greeted us with the Muslim greeting "Aslameleykum."

My relatives greeted them back, "Welekumsalam." So far so good. We all had expressed our desire for peace, and it was more than just a formality.

Dijwar entered as well, with Checkdar behind him. He greeted the others, shaking their hands one by one. The men sat down and started to talk and laugh, drinking their tea and smoking their cigarettes. There was so much cigarette smoke that there was a cloud in the house. It was a good thing that our houses were made out of concrete, which helped to prevent the smell being stuck inside them.

Before supper was set up, one of the elders from Belizan asked, "Can someone tell us why we are here? What is the special occasion?"

My Uncle Sedeeq was a badass. He looked at the old man staring into his eyes and responded without knowing whether the old man was curious about this gathering or was just talking out of stupidity. "Are you happy to be here now? Because if you are happy then you will be happier after the meal. If you are not happy to be here, then you will not get to eat, and you will ask yourself, why did I ask that silly question?"

The old man shyly said, "Oh don't be silly. Of course I am happy to be here. This is Salih's house. He cannot be replaced. Sitting here in his home is an honor, even though he is not with us. Sitting here makes us feel that he is still with us. God bless him and grant him paradise."

All the men fell silent. Some of them

choked up thinking about my father and his memories.

Uncle Sedeeq said “Ameen” (amen).

All the men in the house echoed “Ameen.”

My mother called to me. It was the perfect moment after a beautiful subject, that could have turned either ugly or sweet, was cut off by an Ameen and food ready to be served. My siblings tried to help me, but they were too young to do much. I set out most of the food properly and some of my friends from the village whom I had invited joined me to help me set the food on for the men to eat.

After we set the food up, I told my uncles that the food was ready. They replied back, “Keremkin” (welcome). All the men circled around the food. We had various Kurdish dishes; iprax (dolma), kutilks (rounded stuffed meatball kind of rolls boiled in soup), birinc (rice), various soups (some all-vegetable and others with meat), Kurdish naan bread, zelate (salad), teshrib (broiled ribs on naan), tikes (shish kebab), and lots of other extra meat. We didn’t eat like that every day. Having this feast was special, not because we were all going to eat plenty, but because it was a special day in my life. I would be given permission to marry Dijwar's daughter… if all went well.

Dijwar was sitting next to Uncle Sedeeq. It seemed like Uncle Sedeeq had invited Dijwar to sit next to him to start the conversation after the meal. After the men got done eating the meal, they said "Xode zedekit" (God make it plentiful). My uncles replied to the other villagers, "Hewe afyet bit" (we hope it was enjoyable).

After the meal, my friends and I served tea to the men at my house. I knew that when the tea was served, the conversation would start. Uncle Sedeeq would open the subject up with the rest of the men and ask Dijwar directly what his intentions were.

I needed an escape route. I needed to get away. I could not be here. I was not only feeling shy but getting so nervous that I was about to shake. I did not fear anything, it's just that this was a strange experience that I had not experienced before. I had never dealt with or been in a situation like this.

I told my friends to come get me when the conversation was over. I had told them earlier what the dinner and gathering was about. At first they made fun of me, but they were happy for me and in full support, helping out any way they could.

I walked by Beri's house, which was north of the mosque. It was about ten houses away from mine. I knew that both her dad and mom were at my house. Her father was with the men and her mother was with my mother and aunts.

As I got closer to Beri's house, she was sitting on the front porch with her siblings. Her older brothers were not home, just her younger sisters, whom I believe she was watching. I walked right up to her, not worrying about anything around me. Any villager could see me and gossip in the village. It was too late though. The damage had already been done. It was either going to be a yes or no from her father.

I said, "Silav cana min" (hello, my beautiful).

Beri smiled and replied, "Silav, cano" (hello, handsome). Then her face turned more serious. "Why are my parents at your house? What is going on? Did something happen?"

"Don't worry about it."

"Tell me, please."

I took a breath and said "Beri, remember what I told you before I became a Peshmerga?"

She looked nervous. "What do you mean? You told me many things."

"Beri, I am talking about asking for your hand. Do you remember?"

She said quietly, "Yes, yes, I remember."

I said, "Well Beri, today is our day. My uncles and the other villagers are going to try to convince your parents to let us get married."

Beri stood up. "Shivan, stop lying."

"I am serious," I told her.

She had tears in her eyes. As they started coming down her cheek, she turned and went inside her house. I did not want to bother Beri. I didn't know what Beri's tears were for... were they happy tears or sad tears? Was she afraid to get married so young? Was she nervous, like me? Or had she changed her mind? No, it was not that. Had her father already made other plans for her, the kind that could not be changed?

I walked back to the mosque and

waited there until my friends would come and tell me the news. Beri's father Dijwar would say either yes or no. Either way, sooner or later, I would know. I took my pack of cigarettes out and smoked one nervously.

As I was about to finish my cigarette staring at my house from the mosque, I saw my two friends come out of my house. As soon as they put their shoes on, they came running towards the mosque. It was getting deeply dark out and they could barely see me. They called my name, "Shivan, Shivan!"

I said, "Over here."

They came to me and panted, "He said, yes!"

"What? You're kidding me!"

"No, really, Shivan. Dijwar said yes, but he was going to ask his daughter first."

When is he going to do that? Dijwar and all the men are still...

"They are sending the women back to their house now to ask Beri."

I did not know what to do. Should I stay by the mosque and wait for the ladies to walk by or go back to my house and sit patiently until they came back? I was nervous, but the best thing was to go back to my house. To show myself mature, I would stay silent, which was the proper thing to do, being a gentleman.

I asked my friends when was the last time the men had drunk tea?

They said, "Not since the end of dinner."

"Perfect. Get the tea kettles ready. We are going to serve them tea again, so they don't suspect that I left the house."

My friends hurried to my house, grabbed the tea kettles from inside and brought them out to me. I told them to put more wood in the firepit, so we could make the tea faster. I went to the water station by the mosque to clean and fill the kettles with water. As I was pouring water into them, my mom, aunts, and the other ladies greeted me.

"Hello Shivan. It looks like today is your day!"

I did not say anything at first. I just smiled. Then quietly, playing dumb as if I had no relationship with Beri, I said, "Well I am asking for someone's hand, but they have to like me first and say yes before anything."

Some of the ladies said "Ew ji raste" (that is true).

I walked back to my house. The fire was ready. I poured tea into the kettles and waited for it to steep properly. I told one of my friends to go into the house and, if the men ask where I was at, to tell them I was preparing tea with my other friend. He agreed and went inside. Everything seemed to be going well. I pulled my pack of cigarettes out again, offered my friend one, and lit one up for myself.

As we were about to finish our cigarettes, I could see the tea kettles steaming up. I handed one tea kettle to my friend and held the other. We had to be really careful, first because the kettles were full of tea and we didn't want to spill a drop because there were so many men at my house. Second, the kettles were really hot. We did not want to burn ourselves.

I went into the storage unit by our front porch and grabbed glasses to pour the tea into. Luckily, my mother knew that there would be a lot of people over, so she had borrowed lots of tea glasses from my aunts. When we had all of the glasses ready and had put sugar into them, the tea was ready to be poured. You could smell the tea in the air. That night was a little chilly, being the end of summer. There was a light breeze, but the tea smell and the warm feeling made it feel like it was still midsummer.

As we brought the tea to the men inside, we greeted them, "Aslameleykum." They looked at us and replied, "Welekumsalam." We needed to greet them so that they would notice us and that we were serving tea.

Uncle Sedeeq said "Aha ew shuleke bashe" (good work).

When we finished serving tea to all the men, my Uncle Sedeeq called to me, "Shivan, come sit here." I tried to sit away from Dijwar; who was sitting to the right of my Uncle Sedeeq. I was trying to sit on the left side of my uncle, so as not to take the place of honor from Dijwar, when Sedeeq looked at me and said, "No, sit right here," right between my uncle and the man who had just agreed to become my father-in-law.

Wow! Could it have been more obvious that I was turning red in front of all the men at my house? I hesitated for a second, then walked right towards Dijwar and sat in between him and my Uncle Sedeeq. They both told me “Bi xer hati” (welcome). I replied back “Xerindar bit” (it’s good to be welcomed).

My Uncle Sedeeq spoke to everyone assembled. “Men, my nephew is here. Don’t let anyone think because his father is gone that he is left alone. We are all like his father and will do anything for him. Shivan, you have asked for the hand of Dijwar’s daughter Beri, right?”

I did not know how to respond. I looked down to the floor and nodded, quietly saying “Bele” (yes).

He continued, “Shivan, all of us here have asked Dijwar to give permission for his daughter to marry you. Dijwar knows you are from a good family. Dijwar also knows the villagers here, and Dijwar knows you. We know that you are a brave young man. You love your land, your culture. You are a Peshmerga, of which there are not many your age. Dijwar has agreed to give his daughter to you. Now what do you think of that?”

I could not stay silent any longer. People knew that I was not that shy. They knew I had the mouth of a lawyer. I needed to speak for myself, and so

I spoke, "My dear uncles, villagers, and father-in-law-to-be..." When I said that, everyone laughed.

I waited until their laughter toned down, then continued, "I want to thank everyone for coming today. My father would have been very proud of all of you. I am sorry my grandfather is not here today, but he knows of all the events that are currently taking place.

"Dijwar, you know my family and we know yours. My father has helped you a lot in the past and I want to continue that. I just want to thank you for giving me permission to marry your daughter. She will be in good hands. For now, we have to wait for your daughter to give us an answer."

All the men looked at me in surprise. They could not believe what I had just said. I overheard some of them whispering, "He sure takes after his father."

Others whispered, "Wow, what a great mind."

Dijwar smiled and continued to drink his tea. My closing words meant he need not reply until the women brought us Beri's answer. Uncle Sedeeq had put me on the spot, but I had given Dijwar a break.

My Uncle Sedeeq patted me on the shoulder and whispered in my ear, "You have made me proud today. I am happy to raise my head high in front of all these men."

About five minutes later, we could hear the ladies clapping their hands outside and singing, "Ki zavaye, ki zavaye? Shivan zavaye!" (who is the groom, who is the groom, Shivan is the groom)

Uncle Sedeeq looked at me and Dijwar and said, "De herin desten ek" (now you can shake hands).

When we had done that, all the men in the house started to line up to shake my hand and Dijwar's. A lot of them were very excited for me and kissed both of my cheeks. Uncle Checkdar couldn't stop kissing me, lightly wiping away his tears. It was official: Beri's father had given us permission to get married. Now we needed to plan a wedding day.

The village men walked out slowly one by one. My Uncle Azad walked out with them. He grabbed his rifle and started shooting in the air. It was okay because all the Peshmerga were back in the village, since it was dark now. They knew there was no trouble because they knew what was happening at my house. Some of the other Peshmerga by the mosque started shooting in the air, which was a sign of happiness and joy. That day, which had made me so nervous, became one of the best days of my life: hearing the good news that Beri and I were getting married.

The next day, Uncle Sedeeq said he had to leave. They had to go to Amediya to the KDP headquarters to see if they were needed anywhere. My mother had packed him some food which consisted of bread, Kurdish yogurt, tahin, raisins, almonds, and water. Before he left that day, Azad, Rojhat, Checkdar and some of our other relatives from Brifka came to my house. We talked and set a date for the wedding. It would be the last week of September. It was now almost the end of August.

We had one month to plan and could send out a word to my grandpa and grandma about the wedding date. They were supposed to return to Brifki and Belizan in September to visit us. Hopefully, this letter would encourage them not to change their minds and insist on coming later. It was not easy getting from one place to another during this time because transportation was rare and the conflict between us and the Iraqi regime made it impossible to travel normally. There were Iraqi checkpoints everywhere and sporadic fighting continued between the Iraqi government and the Kurdish Peshmerga.

In late September, most of the orchard trees would be ready. We needed fruit for sweets, since we were unable to go to Zakho or Duhok often. The best way for us to get our sugar was from the orchard trees and raisins that we picked. In the meantime, we also tried to find ways for people to bring us back supplies if they were traveling to the city. I needed a new set of khaks. I also had to get Beri a new Kurdish dress. Even though Beri's father did not ask for the normal dowry, our family still needed to buy her some jewelry and give her parents some money. We needed to get our supplies quickly, and I didn't waste any time.

After my Uncle Sedeeq left with his Peshmerga, and we said our goodbyes, I asked my neighbors and other villagers,

"Who is going to the city?" I asked my mother for how much money we had saved. I sat down that evening with my mother and wrote down the supplies we needed for the wedding day. When it came to savings, my mother was a professional saver.

When my father was alive and when we had guests, my father felt ashamed if he could not host the guests well with food and other things of this nature. My mother always surprised him. When she would bring things to the guests, my father would smile and would later compliment her. "You are a keeper. I can't believe how proud I am to be your husband. You always make my head stand tall." The same thing happened to me now.

I had some items on the list that we needed from the city. My mother told me, "Don't worry. We have those items in the storage, which brought a sigh of relief. All of the items on the list cost money, and the people who were going to bring it for us expected some kind of reimbursement. That is just how things were. We did not have much money, but because of the things my mother had saved, we realized that we could do the wedding without any issues.

My father had left my mother with money to be able to take care of ourselves for at least ten years. She had held onto the money for safekeeping. My income was from being a Peshmerga and from the small quantity of crops that we sold from time to time. Most of the villagers sold their crops as well. We had a bit of competition in the village when it came to selling crops. There weren't enough consumers around the village if one wanted to make money. The city was the place to do that.

We would have the wedding outside of my house in the street. We would invite all of our relatives from Berwari Bala, including our relatives in Brifka. We would invite family and friends from villages that were close by such as Baze, Keste, Girka, Qumriye, Hirore and Babire. We would also let others know around Berwari Bala who were related to us or close to the family as well as others as far as Begova, Kani Mase, Bamerne and Amediye. Our singers would be locals from the village. My Uncle Sedeeq was going to bring the Peshmerga commanders who had been especially close to my father. Beri's family would invite their relatives as well. We had done all the planning and needed to start preparing for our guests.

It was the middle of September and my grandparents had made it back from Mosul. We were happy to see them. They had brought a lot of supplies back with them. Since they received my letter, they had brought things for the wedding as well. They were deeply excited for me and were happy to hear that someone was going to help my mother around the house. It brought my grandparents comfort. They knew my mother needed a lot of help. My siblings were young and there was a lot of work to do around the house and in the farm fields.

We had gotten all of our supplies for the wedding. There were still two weeks left of September, and our goal was to begin the wedding in the last week of September, starting the Monday of that week. Traditional Kurdish weddings were usually three days long. There was the Xenha Night, the pre-wedding party. Then there was the wedding day, followed by a picnic on the third day with family and close friends.

Our family was going to celebrate the whole week. We had a large family and many friends coming from near and far. We needed more days to celebrate with them, to make the trip worth it for all, because it was a tough travel in terms of terrain when coming to Belizan.

Day by day, we saw our relatives arriving in Belizan. They brought supplies with them and looked festive in their new Kurdish clothing that they were showing off. Friends of our family started coming a couple of days before Monday to stay near our village in Belizan. Everything was becoming real. I was very happy and welcomed our relatives and guests. It was a good thing Belizan was a big village and a lot of families lived there at the time. There was room to host hundreds of people. There were over one hundred families living in Belizan even though some of the families were not from Belizan. There were families that had escaped from Iraqi troops, like one family from Nezdore now living in the Guliya area near Zakho.

That Sunday, Uncle Sedeeq arrived with some of his Peshmerga. This time there were not that many with him. There were usually over thirty Peshmerga with him, but this time there were only ten, all of them elderly men. They did not look like regular young Peshmerga fighters. They looked worn and scarred, with white hair and wrinkles on their faces. I knew that all of these Peshmerga were commanders. I also knew they all had great stories of battles to share.

Each of them came to greet me, kissing me on both of my cheeks. They mentioned that I resembled my father, and that they had been good friends with him. I welcomed them and escorted them to my house. They were the most respected people in the village, and if I invited anyone to my house to stay over, it would be them.

They sat down and I quickly prepared the tea kettles and started the fire to make tea. I knew that more of our relatives and friends would stop by, so I made sure that we had plenty of tea ready. Kurds love tea.

Slowly, more of my relatives came by. I did not have to work as hard anymore.

We had lots of help around the house as relatives and friends started arriving one by one. The wedding was becoming more official each minute and hour that passed.

By Sunday afternoon almost all of our relatives and friends had come to Belizan. You could hear people talking everywhere. At my house, in front of the street, by the mosque. Noise came from other villagers, family, and guests who were everywhere. Even someone not from Belizan would realize that there was a celebration going on.

Some of the ladies, my cousins and friends of theirs, had gone to Beri's house to see what she needed and bring supplies for her to prepare her to look her best on her wedding day. It was traditional for the groom not to see the bride until the wedding day. I had seen Beri in the village, but I had not had a chance to speak with her one-on-one since asking for her hand. Even when I did have a chance to talk to her, I did not. I did not want people to talk bad about us. After all, we were going to marry each other and going to spend the rest of our lives together. "*It's not worth it to bring shame to both of our families. I will keep my distance and hopefully she will do the same and realize why,"* I thought.

Uncle Rojhat had brought a couple of Kurdish Gypsies to the wedding. They would play the duhul (Kurdish drum) and zirna (a flute-like instrument that had great music to it) so that people could dance. At the time, they were popular amongst the villages. He heard that the musicians were in Baze for a wedding and walked there to get them before they left the area. That was his wedding gift to me since he was stingy about giving money away.

It was early Sunday evening. After we had lunch, supper, and lots of tea, the duhul and zirna started to play. I knew that it was time for the pre-wedding party, or Xenha. Kids were clapping and jumping around. The men were dancing on one side, singing, and the women on the other side, singing with them in a duet. I was sitting down with some of the elders from Belizan and my relatives from Brifka.

We heard singing and dancing on the other side of the mosque as well, where Beri lived. We were in two separate locations, celebrating. Back then, we could not see each other until the wedding day. This was out of respect for the bride's family and home.

Some of my relatives where there, keeping Beri company with dance and singing. It was mostly women with Beri, except for some of her family members. Most of the men were at my house dancing, singing, and standing around.

Every couple of minutes, my uncles and cousins would come up to me and ask if I needed anything. When a man or woman got married, they got treated like a king or queen on their day. Everybody was their servant to please them and keep them happy until the wedding was finished.

The next day, everyone was up very early. I thought most of the people would be tired from the night of Xenha, but that was not the case. Most of the villagers were up with their guests and had finished breakfast before eight in the morning. My uncles were busy setting up the wedding space. They had brought a horse that was going to be our transportation to pick up Beri and bring her to the wedding reception area. The horse was decorated in red to match Beri's dress. Red was a symbol of being pure for the woman getting married, that she had not lost her virginity yet and that she was the first for the man to marry. There were other cases of course where divorced women or widows ones got married. They did not wear red though. They wore traditional Kurdish clothing most of the time and had smaller weddings compared to someone who had never been married before.

My wedding was big, since it was our first time, and there were over five hundred people there. There were local villagers from Belizan, our neighboring villages nearby, friends of my fathers, Peshmerga friends of mine, and my relatives and family from Brifka.

Everywhere I looked, someone was doing something. There were about ten ladies who were making lunch for all the wedding guests. There were other young girls washing and cutting vegetables. The men brought wood from all sides of the village, and my cousins my age were bringing water and fruit from the farm fields. The only people who were not doing anything were the Kurdish commanders, my Uncle Sedeeq, and some of the older men from Belizan and Brifka. They were the respected ones and were excused from doing work since they were the important guests of the wedding. Even if they wanted to help out, others would not let them. That would be very shameful and not how one treats their respected elderly guests in the Kurdish culture.

The afternoon came and Duhul and Zirna came out again and started to play their instruments. The women, men, and kids started clapping. With my uncles Azad, Rojhat, Checkdar, and Sedeeq, along with other relatives, cousins, and friends by my side, I went into a private room to take a shower. They had my Kurdish clothes ready when I was done.

I was going to wear bergiz, the more upscale version of khaks. They were my father's. They were blue in color, which was his favorite, along with a white, long-sleeve shirt underneath. All my relatives, local villagers, guests, friends and cousins got me ready. They dressed me up and put cologne on me to smell good. Each of them came and shook my hand, hugged me, and some kissed me on my cheeks and forehead.

After they walked me in front of my porch, my uncles asked my mother, "Aysha are you ready?" My mother was inside. She was trying to hold her tears back, which she could not, and so she ran back into the house. Some of the ladies went inside with her to comfort her. Finally, she came out. She shouted, "Tleeleee"! (a sound made by the mouth for joy and excitement), and then other women followed, "Tleeleee!" They put me on the horse, and we headed for Beri's house to pick her up.

We arrived at Beri's house with hundreds of people who came with us. Her father, her mother Zubaida, her brothers, and other relatives were on the front porch. Some of them were smoking cigarettes and drinking tea. When they saw us, they all got up quickly. Beri was inside of the house.

The men took me off the horse. I started walking towards Dijwar and my close family walked behind me. The rest of the people kept their distance. One by one, we shook their hands, with Dijwar being the first. Dijwar said, "Hin bixer hatin" (you are all welcomed here).

The ladies went inside to get Beri. They brought her out. I could not see her face because a veil covered it. They brought her by the horse, and my uncles and I helped her onto it. We shook Dijwar's hand again, along with his sons and his relatives that were there. Then we excused ourselves and headed to the wedding reception area, which was in front of my house.

As we walked back, my mother got a jar full of candies and threw them in the air for the kids to catch. They got excited and started running in all directions toward the candy.

The wedding started around three o'clock that afternoon. I lifted Beri's veil off when we arrived and before she sat down. She looked so beautiful that day, more beautiful than I had ever seen her before. She looked at me, smiled and sat down.

Our wedding feast lasted until the early evening. People danced, sang, ate, then sang and danced again until it was dark. We did not have electricity and relied on lanterns, candles, and fire for light. My house was left empty for me that night. My mother and other siblings were going to spend the night at Uncle Azad's house.

After the wedding party, Beri and I were escorted to my house. We could hear the wedding guests up and still talking and laughing around the village, but we were given the rare gift of privacy. Beri and I did not waste one second. We hugged each other, kissed each other all over, and made love, real love that we had not made before. Beri and I were finally married the fall of 1986.

CHAPTER 9

All-Out War

It was winter of 1987. Beri and I had been married for a little over a year. Things were going well for us. Beri was helping my mother around the house. While my mother was in the fields farming the crops, I would help both of them when I was not gone on Peshmerga campaigns. I would leave once a month for about a week and then come back.

I had been a Peshmerga for three years now. When on duty, I was stationed at various villages in the outskirts of Duhok City area. There was not much activity from the Iraqi soldiers. During this time, Saddam Hussein was not only fighting the Kurdish uprising in the Kurdistan

Region of Iraq but was at war with Iran as well.

When we received news, we would hear of the thousands of soldiers being killed on both sides. We would hear how young men, Arab and Kurds alike, were forced by Saddam's Ba'ath Regime to enlist in the military to fight. With little training by either the Iraqi or Iranian regimes, the forced enlistees were easy targets.

As months went by, we heard that both armies were going against International Law and were using prohibited weapons in war. Things started escalating after an all-out war was finally declared by Saddam Hussein.

Near the end of 1987, Saddam ordered his top generals to squash the Peshmerga resistance in the Kurdistan Region of Iraq at all costs. He was lying to the Iraqi people and his military that the Peshmerga were the ones to blame for losing to Iran.

We got orders from the KDP headquarters in February of 1988 to prepare for the worst to come: we were going into battle against heavy weapons from the Iraqi army, such as tanks and jet planes.

In our minds we knew that we were no match for the Iraqi army. Our AK-47 rifles and small weapons didn't stand a chance against them. We were going to be outnumbered and outgunned.

The Peshmerga had to use all the skills at their disposal to defend the Kurdistan region from the new attacks coming from the Iraqi army. Our advantage was the mountains. The Iraqi troops were no match for the geographic terrain in Kurdistan. They only had an advantage when they knew our positions and bombed the specific bases or bunkers.

It was the middle of March. Spring was coming. We were supposed to be celebrating Newroz soon, the Spring Festival. News came that all enlisted Peshmerga had to go out and fight off the Iraqi occupation. The Iraqi regime was in almost all the Kurdish cities, towns and villages. Kurdish civilians were being killed left and right. We needed to go and save as many of our people as we could.

Once again, the time came to say goodbye. My mother, Beri, and siblings held me tight. They did not want me to go. They knew this time I would be gone for a while: maybe three months, six months, or God forbid, forever.

When times like this came, which were often, it was hard to hold back tears. Our struggle in fighting for freedom felt hopeless when a change was nowhere to be seen. It seemed like the fighting was continuous. Just when we thought peace was coming another tragedy would occur and turn our happiness to grief.

My Uncle Azad was getting ready too. At that time, we were the only two that were Peshmerga from Belizan village. There were other Peshmerga from other villages who were staying in Belizan but were not from there. We met by the mosque, said our goodbyes to our family, and headed to Amediya where the KDP headquarters was. All the Peshmerga were to gather there for assignments to various places throughout the Kurdistan Region of Iraq.

We left early in the morning and arrived in Amediye in late afternoon. We were more cautious going there, which is why it took us so long to get to the KDP headquarters branch there. Jet planes flew over us along with helicopters that we could see at a distance. Iraqi military vehicles came and went. It was a frightening situation to be in.

Before sunset, we managed to get to our destination safely. Amediya was breathtaking when we arrived. The beauty of the waterfalls and Kurdistan's natural beauty made me appreciate what I was fighting for. The people who greeted and welcomed us made me feel proud that I joined the Peshmerga to free them. I was scared deep inside but relieved that if something were to happen to me, my legacy would not be forgotten. I would be remembered as a shehid, a martyr who fought for his people's honor and homeland.

In Amediya there were Peshmerga everywhere; thousands of them gathered. Everywhere one looked, Peshmerga were stationed near the headquarters. My uncle and I went to check in at the base. We were greeted by a dozen other Peshmerga who were guards at the door. They asked us a few questions and then welcomed us into the station to get checked in. There, Azad and I saw our other relatives from Brifka. They came to us, hugged us, and gave us kisses on the cheeks. I felt comforted that all us were there together.

About a dozen of us blood-related men met there. We looked out for one another while we stayed at the headquarters making us feel safe. We knew, though, that all of us were going to be separated soon because of unit assignments. We spent as much time as we could together until we received those assignments.

There I met Ismet Hasan Tirwanishi. He was sitting at a fire pit making tea. I sat next to him and introduced myself. He shook my hand and welcomed me. He offered me a tea and we started to chat. We connected very quickly. That's when he told me his story.

Ismet told me that he became a Peshmerga at the age of twenty-five. "I went to the Bamerne airport in 1974 to get my weapons. I was supposed to go see Selim Eshed who was in charge of giving the Peshmerga weapons. I had not been accepted as a Peshmerga yet but had enlisted. I, along with a few other Peshmerga volunteers, were told, "Sorry but there are no weapons now. We will send word for you when some become available.

"There was six to seven people who

were with me, and they were all from the same village as I was, Tirwanishe. We went to Bigeele Village together. There was a force there called Hiza Himreen (Himreen Force). We knew someone with the unit called Areef Hasan. He was from Girka village, in Berwari Bala. He oversaw recruitments for the Peshmerga Battalion. We went to visit him, and he knew who we were. He welcomed us.

"Areef said, 'To be honest with you, we don't have any weapons available to give to any new Peshmerga. However, we will write down your names. When we get the weapons, we will call for you and then you can serve as a Peshmerga after that.'"

Ismet continued, "We spent nine to ten days there in Bigeele Village. They wrote our names down, and then we returned home to our village. A month or two passed, and we received a letter to go and get our weapons. The fight had started between us and the Iraqi government again. I and the other Peshmerga from my village went to pick up our weapons. We were each given a single-shot Brno Rifle. There were automatic AK-47s, but the quantity given to the Peshmerga was very low. We stayed with our unit for five to six months. We were then given a vacation to go see our families.

"Our Peshmerga Battalion then was very large. We served with Areef Hasan, the officer from Girka Village who had sent us home the first time. There were a lot of Peshmerga from the Berwari Bala Area. There were some Brifkis with us as well. I came home for my two-month leave. The situation between the Kurds and the Iraqi government got worse beginning the third month I was home, just before I was reporting back to Hiza Himreen.

"Our leader, Mustafa Barzani, called all the Peshmerga to retreat to the Kurdistan Region of Iran in 1975. We were forced to leave the Kurdistan Region of Iraq immediately. Our leader, Mustafa Barzani, told us that those who could surrender to the Iraqi government should do so if they knew that their life was not in danger. This was a sad situation for the Kurds, especially the Peshmerga.

"Many Peshmerga left for the Kurdistan Region of Iran. Many others surrendered to the Iraqi government, along with their weapons. Many threw their guns down and acted like they never were part of the Peshmerga. For myself, I stayed at my village until 1977. That is when the Iraqi government came.

"They removed us from our villages. We were ordered to live in or near cities. No Kurd was supposed to live in a mountain village any longer. My family and I were relocated to Bersive, near Zakho city. We lived there until the beginning of 1980.

"We got word that the Peshmerga were back in the Kurdistan Region of Iraq. This was great news. I headed back to my village, Tirwanishe, once I received the news. Once again, I joined the Peshmerga in 1981.

"When I went back to the Berwari Bala area, I enlisted in the 1st Regiment Peshmerga Unit headquartered in Amediya. When I joined the Peshmerga again, I was sent with Ahmed Oray and his unit. We were around three hundred Peshmerga. When the Iraqi government knew that the Peshmerga units were getting larger and stronger, they would send leaflets and messages to the Peshmerga. They would offer them money and a mere slap on the wrist if they withdrew from being a Peshmerga. Because of the living situation and the Peshmerga's families, many did leave our units and surrender to the Iraqi government. I stayed as a Peshmerga at the headquarters of Amediya until now."

I looked at Ismet with respect. "That is quite a story. I joined three years ago. I don't regret it one bit," I told him.

He looked at me and smiled. He poured another tea and asked if I wanted another one.

I said, "Yes, but I will pour the tea into the cup myself."

He shoved my hand away and said, "Ma chiye" (it's not a big deal). He asked me, "Whose battalion are you with?"

"My Uncle Sedeeq's."

He smiled and said, "Sedeeq, he's well known around here." He continued telling his story while drinking his tea. "Most of my time as a Peshmerga was in the Berwari area, from 1981 to now. I had gone to the Kurdistan Region of Iran twice. One time, I was with my Peshmerga unit when we were ordered to go. We were stationed in Berkeme, where there was another KDP headquarters branch. The commanders there were

Fazil Mitni, another commander named Ezedin, and other KDP commanders were there too. There was a big KDP base there and we were to guard it.

"There was a fight between our party and another Kurdish party there called the Patriotic Union of Kurdistan or PUK in 1983. The good thing was, it didn't last long. It was only for a few months. I was stationed there for two years.

"Another time I went to Iran to see a doctor there. I was in the Kurdish area. I had treatment for about two months. That is how I had seen the Kurdistan Region of Iran. After my two years were over, I was relocated back to Amediya in the Kurdistan Region of Iraq. I have been at the Amediya KDP base ever since.

"When I became a Peshmerga, I was a volunteer, so I bought my own rifle and ammunition. The KDP did not have much military equipment. They only received it when there was an ambush or a successful attack on the Iraqi soldiers. Sometimes, if the KDP was lucky, they would get weapons from the Kurdistan Region of Syria, where there was a base as well. The other times that weapons were received was if they were smuggled into the Kurdistan Region of Iraq and handed to the KDP Peshmerga or sold to them.

"I bought my rifle for about one thousand two hundred dinars from an arms smuggler. I had my own rifle with me until I was called to serve in the Kurdistan Region of Iran. That is when I sold it to support my family, because I knew I would be provided another one by the KDP. They gave me a Wasat weapon, similar to a Brno. I traded that in for a Seminoff and carried that for about a year.

"In the early months of 1987 we had our big break. We attacked an Iraqi military base and retrieved tons of weapons including various automatic rifles. I got one called a Sikho.

"The only time I was paid right as a Peshmerga was during the Shorash Eylun time from 1974 to 1975. Since then, I have been paid ten dinars a month. From 1981 to now. Sometimes we get more, but not often. We were not given a monthly salary. The things we got during this time was usually clothing donations such as shirts, shoes, pants, and socks. There was no money to be given to the Peshmerga by the KDP. From 1981 to now, many months went by where I was not paid by the KDP. It was all volunteer service.

"There were many attacks on the Iraqi soldiers that I was a part of. One time we attacked the Bamerne Iraqi Army military air base in 1985, which was close to Duhok City. There, we took lots of weapons. We also took military Humvees and jeeps. Unfortunately, we lost a Peshmerga named Fowzie. He was from your village, Brifka. His father's name was Remezan and he was his only child. His father Remezan came to us one day and told my commander, Ahmed Chelki, 'Please look after my son.' We attacked the Bamerne military air base the next day and that is where he was killed.

"Another big fight I was in was at Tine village, close to Bamerne town. There, we lost another three of our Peshmerga and many other of our Peshmerga were wounded. All of the three Peshmerga were from Berwari Bala area. One of them again was from Brifka village. He was a handsome, well-built man. Another Peshmerga that was killed during the fight was from Ormadade village. The last Peshmerga, named Salam, was from Derishke village. Those three Peshmerga we lost that day were all family men.

"There was thirty of us Peshmerga that had attacked the Iraqi base there. Out of the thirty, we had lost three, and twenty-four others were injured during the fight. We had such heavy losses because they placed mines all around their base. We sent scouts a couple of days before. They thought that they had picked up all the mines, but they were wrong. When we attacked some of the Peshmerga stepped onto the mines and they were blown up.

"The Iraqi troops fired at us with machine guns and grenades. We had to retreat and that is when we lost three of our Peshmerga, each of them shot in the back of their heads. Luckily, little by little, we were able to take our dead and the injured. We made sure that no men were left behind. We lost that day.

"A year had gone by and our unit stayed in that same area. We did not fight until we got more reinforcements; then the Peshmerga attacked again. I was ordered to stay in Amediya as a guard, but we got word that the Peshmerga had taken over the base that we had lost. We were very happy with the news because our martyrs were avenged. There were about fifty Iraqi soldiers in the base. The Peshmerga had killed most of them but also took prisoners."

He looked at me and asked, "Am I boring you?"

"No, thank you for the tea. Please go on. I am enjoying your company."

Ismet kept on talking. I drank it in.

At last we heard, "Alright, everyone to their units."

A few commanders kept repeating the order until each Peshmerga headed to their unit. I shook Ismet's hand and told him I would see him later. He said "Bele" (yes), we will see each other soon.

That night those of us who were not guards slept early. In the morning we were to go to Kani Mase, the capital of the Berwari Bala area, where thousands of Iraqi soldiers had been stationed. At first light we were all to set out for one of the biggest battles that we had ever fought. This was going to be a fight the Peshmerga had never seen before.

There were now three Iraqi brigades

stationed there. One we called Jehshik, which was full of Kurdish traitors who served in the Iraqi army. The other two Iraqi brigades were from other Arab cities, and all three together consisted of thousands of Iraqi troops.

We woke and mustered around five in the morning. There were three Peshmerga companies assigned to attack. One came from where we were at in Amediya, one came from Duhok, and the other from Zakho. Together they were over one thousand Peshmerga strong.

We had breakfast. Some of the Peshmerga started to dance and others joined. One unit after the other lined up and headed for Kani Mase in Berwari Bala.

Our orders were to stealthily surround the Iraqi brigades and launch a surprise attack at twenty-three hours, or eleven PM.

We surrounded the Iraqi brigades without incident. Through the binoculars, you could see Peshmerga quietly setting up their positions to attack when the time came. The orders were that the Peshmerga all had to attack at once. I was nervous. I had been in battles before, but not one this big. There were tanks that I could see at a distance, machine guns on bunkers and buildings, and thousands of soldiers who were talking out loud in Arabic. We were close enough that their voices drifted to us on the evening breeze. It was an intense moment in my life.

Nightfall came and we knew that the fight was getting closer and closer. Our bodies were in adrenaline mode. My Uncle Sedeeq told us stories of the brave from the past, to calm us down.

Bang! One shot from the Peshmerga side. For a few seconds, there was calm. Then after, I could hear gun shots going off from all directions. My Uncle Sedeeq looked at us and said, "Ek, du, se (One, two, three) and then we poured our thousands of bullets all at once and rushed for the Iraqi soldiers. We fought all night to secure their base; the battle lasted until morning light.

When morning came, we had full control over the Iraqi brigades. We had killed thousands of their soldiers. Just a few hundred were left, whom we took as prisoners.

There were two Iraqi bases remaining

in Sere Kore Zeri, in Bewari Bala. We attacked and overwhelmed those bases a couple of days later. The Peshmerga liberated the whole Berwari Bala area and were advancing in other Kurdish areas of Duhok and Zakho.

We made Kani Mase our new base. We had been there for several days when I ran into Ismet Tirwanishi once again. We sat down together at one of the cafés there.

He asked, "Have you heard?"

"No, heard what?"

"There was word from the Zakho Peshmerga base that we were going to be bombed with chemical weapons. We got word from the highest KDP headquarters to let the Iraqi troops back into the areas that we had just liberated. If we didn't let them return to their bases, we would be gassed."

I looked at Ismet and we were both sad because we had lost many Peshmerga friends, some who were very close to us. He said, "I have lost so many good friends. Saeed Chelki was one of them. He and a few other Peshmerga were scouts. They were out removing land mines near an Iraqi base to prepare for a Peshmerga attack, and one blew up on him. He died instantly."

I looked at Ismet and said, "Xode je razibit" (God forgive him).

"Ameen," he intoned, and continued, "Xisro was another person close to me who was martyred. He was injured badly in Derelok village after attacking an Iraqi base. The Peshmerga had brought him back and were treating his wounds but could not save him. He died a few days later. He was also from the Berwari Bala area.

"A few other Peshmerga in my unit, from Chelke village in Berwari Bala had been killed. There was an Iraqi base in Girka village. Chetes were there, Kurdish traitors who joined the Iraqi army. We fired a few rounds at them. They yelled, 'We are Kurds! We surrender to you, please don't shoot.' We decided not to shoot them." Ismet took a sip from his tea and lit a cigarette. Then he sighed and said, "We took them as prisoners, but on our way out of the base, one of our young Peshmerga, named Shukree, who was also from Chelke village, stepped on a mine and was killed instantly."

I shook my head. "Xode je hemiya razibit" (may God forgive them all).

"Ameen. I can't believe that after all the Peshmerga lives lost, all the Peshmerga blood spilled, now we have to give up all the areas that we have won."

I told him, "We did our job. They know not to mess with our people. Maybe this will be a sign for them to give us our rights because they cannot beat us on the battlefield."

He looked to me and said, "I hope you are right. You have a great mind. Well, I must go to take care of a few things before we are forced out of here."

I got up, gave Ismet a hug, and watched him disappear into the crowd of Peshmerga.

I sat down at the café. One of my cousins in my unit came by and asked if I needed anything. I told him I was fine and ordered another tea. I started thinking. During my time as a Peshmerga, I had gradually begun to give our civilians more credit than us Peshmerga. They had to go through a lot to protect and serve us. The villages, left untouched by the Iraqi army because they were farmers, helped us a lot.

The Peshmerga relied on the civilians to stay alive. The Peshmerga were not receiving supplies from anywhere. Most of them were volunteers. Throughout my time as a Peshmerga, our people did a lot. They fed us, gave us shelter, provided clothing, and gave us information on the enemy. The closest Peshmerga branch was in Amediya. If a Peshmerga was lucky to be near there or stationed there, they got all these things. Otherwise, in other areas, everything was scarce for us. The only people that we could depend on were the villagers. In the villages is where we were fed and clothed, and even cared for when we were ill.

There were many families from local villages that had a great reputation amongst the Peshmerga, families who treated us very well. They made sure that we had all of the supplies we needed to do our duty. In Berwari Bala there were many good villagers that helped us.

I remember Ismet talking about one of them. He had told me, "One person who helped us a lot that comes to my mind is Haji Salih. He was from Kani Mezin village. His house was like a Peshmerga base. There were dozens of Peshmerga at his house all the time." As I remembered him saying that. I too remembered all the families who had helped us. Especially my family, my own father and mother, who had given the Peshmerga so much food, shelter and protection.

I could not believe what I heard, that we had to retreat from the bases and villages we had liberated. We were going to be letting our people down once again. From 1981 to now, a lot of village folks from Berwari Bala helped the Peshmerga. We had much appreciation for all of them. Now they would have to deal with the Iraqi soldiers once again after we left.

Our people all had gone above and beyond in serving us. The happiest moments as a Peshmerga for all of us were before and after we attacked the Iraqi bases. Before, we would sing, dance together, and then grab our rifles and attack the bases. After taking over the bases and bunkers, especially when we did not have any causalities, it was a great moment of pride and happiness. Even when we lost our fellow Peshmerga, we understood that was the cost of victory.

We all knew what the name "Peshmerga" meant. We were facing death, and that was what a Peshmerga was. We had much pride, though. We did not care if we were martyred. We were fighting for our homeland. If we lost our lives, it was because we fought for freedom with our heads held high.

Before attacking the Iraqi army, our commanders would tell us to sit down with them and they would explain what we were going to do. If a Peshmerga was not selected to partake in an attack, they would easily get upset. Everyone had so much pride and didn't fear death. We were all fighting for a free homeland. All the Peshmerga who were selected were so happy that they would have that chance to attack the Iraqi army. It was a sign of the bravery that was in all of our hearts.

When we did lose Peshmerga, we would send them back to their villages and towns to be buried. If their families were in Duhok or Zakho city, we would send word out to allow us to send their bodies there after their families had gotten permission from the Iraqi government, which had a lot of Kurds working in it. The connections were there, and we made sure all of our martyred Peshmerga had proper burials.

The sad part of being a Peshmerga was listening to the stories of locals who were harassed by the Iraqi troops when they would bring food and supplies from the city back to their villages and towns. The Iraqi troops had checkpoints in and out of the cities of Zakho and Duhok. They would take the food that the locals brought with them from the city and would throw it away. They accused many of them of taking food for the Peshmerga. Many of their families would go months without food supplies. They harassed the locals and called them dirty names. They spit on them and swore at them.

When the locals returned to their villages, we would see the sadness in their faces. We felt their pain and the anger inside of us grew bigger and all we could do was take it in. I don't think any of us regret being a Peshmerga. We did it for the Kurdish people. We did it for Kurdistan. The elder Peshmerga fighters would say, "if we could go back, we would do it again. If we could serve as a Peshmerga like our young days, we would still be doing it." Some of them wished that they had been martyred as a Peshmerga because they would always know that they had died for a just cause.

CHAPTER 10

Forced to Leave

It was Tuesday, August 2nd, 1988, my birthday. We did not celebrate birthdays; no one in the village really cared much about them. I knew that in the city, people did, though. I knew that it was my birthday and I told myself that I needed to do something fun, but I did not know what. I was now eighteen years old. There was still a lot of fighting going on between us Peshmerga and the Iraqi army. I had lost a few good friends who had been close to me throughout my years as a Peshmerga.

I was happy because it was my birthday. I was also happy because Beri and I had welcomed our first child into the world two months ago this day. He had been born on Tuesday, June 7, 1988. We named him Welat, meaning "Homeland." We did not know the gender of the baby before it was born. There were no hospitals or clinics nearby that could have told us the gender of the baby.

We relied on a midwife to deliver our first child. It was very painful for Berivan. I had heard her screaming and crying while giving birth. I had stood near her, just outside of the birthplace, but did not want to intrude on the women's privacy. Fearing that I might lose Beri or the baby at any moment, my heart raced and I sweated throughout my body. Luckily, my uncles were by my side. They kept me talking and helped take my mind off anything bad that might have happened.

My mother had walked out to us and said, "Shivan, God has given you a son." I had smiled as she walked back into the house.

I had waited a couple of minutes outside the birthplace. Finally I had knocked on the door and gone in. I had been so happy when I saw the child. I had tears in my eyes. I had been scared to hold the baby and said, "I am happy, Mother.

Thank you, God. I will be in the room in a minute."

My uncles heard my mother's announcement and came to me. They shook my hand and congratulated me. They said, "Jiye we ye direj bit" (may he live a long life).

The year 1988 started out with lots of fighting. I was still a Peshmerga. I was gone after the new year for about a month. I came back home beaten up and tired. We had had a lot of battles, attacking Iraqi bases in Begova, Bamerne, and the biggest one in Kani Mase, Berwari Bala. This had been a horrific year when it came to fighting.

Saddam Hussein, the dictator of Iraq, was going all out on us Kurds in the Kurdistan Region of Iraq. He was also fighting with Iran. He had gassed the Kurdish city of Halabja on March 16th, killing thousands of innocent Kurdish civilians, and blaming the Kurds for his losses to Iran in that area because the city of Halabja bordered with it.

Kurdish cities were under strict curfews during the night and any suspicious person— male, female, young, old—were to be arrested, and even executed, without trial the same day. Imagine how it was in the city, which was probably worse than in our villages. At least in the villages we had the mountains to protect us.

There were constant planes flying around, helicopters above our heads, and more Iraqi troops being sent to the Kurdish region. There was something planned by Saddam Hussein's regime, but we did not know yet what his intentions were. We knew that a Kurdish massacre must be taking place in the bigger cities because we had never seen so many Kurds from the city fleeing to the villages.

His operation was called "Al Anfal" (spoils of war), a verse he used from the text in the Islamic book, the Quran. Using the Islamic religion of the Arabs was another tactic of his, to cast the Kurds as infidels and non-believers, to claim that we were children of the Fire and not to be trusted by Muslim Arabs. Saddam proclaimed that killing us was not a sin.

There was no place that the Kurds

could hide, unless one was a traitor, a Kurd who hated himself. Then they would be accepted to some degree, out of necessity, but always regarded with suspicion and prejudice by the Arabs. Otherwise, even though we supposedly had more rights in Iraq as Kurds than other Kurds in Iran, Syria and Turkey, those rights were rarely enforced by the Iraqi government. Always, among Arabic Iraqis, we felt like outsiders. We were being forced out of our homeland little by little.

Many Kurds living in Arab cities didn't dare to speak nor dress Kurdish, but chose to blend in with the Arab population. They did this to avoid prejudice, threats, vandalism, or physical harm to themselves and their families.

The sun was becoming brighter and brighter. I knew that afternoon was coming. It was a hot summer day. For my birthday, I was going to go to Kani Teziye for a quick dip in the water. The water was so cold that it shivered your body once you dipped into it, no matter how hot it was outside. Beri was home with my mother feeding my son.

I was enjoying myself until I heard loud screams from the village.

I could hear lots of footsteps, like a crowd of people running. I heard rocks rolling fast down the road and hills. I knew something bad had happened. I quickly got dressed and headed for the village. I came to the village and there was panic everywhere.

The villagers were packing supplies and heading north out of the village. I asked a few of the villagers what the big fuss was about.

They replied, "They are coming. The Iraqi soldiers are coming!"

I saw a group of Peshmerga near the mosque. I went to them. "Xere bra?" (what is going on, brothers?)

"We have received intel that we need to evacuate all the villages. Saddam has a one-hundred-thousand-strong army heading to these villages to wipe us all out. Their orders are to imprison or kill anything that moves. On top of this, he has ordered deadly chemical gas to be dropped in areas where the Iraqi troops can't reach."

I was stunned. "Are you sure about this?"

The Peshmerga replied, "Bra (brother), we are here to pass the message and be guards until everyone is able to escape safely."

I said, "Sopas" (thank you) and ran to my house.

My mother and Beri were already packing. My mother looked at me and said, "Shivan, did you hear? They are coming for us, Shivan! They are coming for us!"

Beri said "Shivan, what should I bring?"

I told them to pack just a few days of food supplies and some things to sleep on. Maybe we would be back soon after the raids.

They quickly packed food supplies, blankets, and pillows. I grabbed my rifle, pistol, grenades, and put my ammunition belt full of bullets all over my body. I was ready just in case we could not make it, or faced fighting along the way. I took my son in my arms. My siblings walked on ahead, followed by my mother and Beri. I walked behind them, and we started walking towards the Kurdistan Region of Turkey's border.

As we walked, we could see thousands

of people ahead of us and behind us going towards the Kurdistan Region of Turkey's border. People were running out of their villages. You could hear kids crying, men in panic, women confused, the elderly barely walking, and small children crying in their siblings' and parents' arms.

One could tell that everyone had packed up in a hurry because they were not carrying much with them. They knew that they couldn't bring much with them. One person next to another and one family beside another, they started forming a line and heading northeast towards Kani Mase, the capital of Berwari Bala; the place we had taken over from the Iraqi soldiers almost a year ago.

No one knew their fate.

The message had been to leave, to get out as fast as possible. That is exactly what we did. As soon as we got the message, we packed as much as we could and left. We did not want to be helpless victims like the people of Halabja. The Peshmerga had warned them to evacuate the city as well, and they had not evacuated. When Saddam's planes came, they dropped lethal chemical bombs, releasing gas that killed its victims instantly. Thousands tried to escape at the last minute, as the bombs dropped, but they didn't get far.

We heard about what had happened. We did not see images nor watch it on TV. We did not have access to satellite TV. The only news information we received was by radio, by word of mouth from relatives in the city, or by the Peshmerga. Half of the time we could not get a good signal from the radio, but when we did, that is how we got most of our news. Getting news from the Peshmerga wasn't very consistent either, because most of them would be away for months and their news was old. Unless it was from the KDP branch in Amediya.

The walk was long and seemed to take forever. We could not walk fast either. We were walking behind other people. The line ahead of us was as endless as the one behind us. I was carrying my son Welat most of the way because he was in his crib. I would sometimes carry my little brother Heja, the youngest of the family, on my shoulders. He kept on crying that he was tired of walking. I finally gave the baby crib to Beri and said, "I am going to carry Heja." She agreed and I put him on my shoulders.

After a few minutes, he stopped crying. He felt safe on my shoulders. He was calm. Sometimes my right arm would fall asleep because I was carrying Heja, and my rifle draped around my shoulders was too much. I had my other ammunition around my waistline. I was carrying about forty extra pounds on my body. I was sweating like crazy, but just kept on walking.

I did not see anyone stopping. People just kept on walking: young and old alike. The sun was about to set by the time we reached Ore village. We had not reached the Kurdistan Region of Turkey's border yet, but we needed to rest before nightfall. People hurried to grab a spot to sleep for the night. The people of Ore had left their village already. There wasn't one single person left there, just farm animals. It was too dangerous for us to stay in the village. The people with whom we had fled camped with us on the outskirts of the village. It was good to be on the outskirts. We would act like trees in the morning if we saw planes above our heads. We had no tents, our food supplies were low. People were asking each other for various supplies. We needed to rest for the night for our journey to the border was still a half a day more. Having children and the elderly made the situation tougher as well.

That night, I looked into the sky. It was full of stars. I wanted to talk to God, ask him why we were going through this. My heart was in pain. Not for me, but for my family. Where was I taking them? What is our future going to look like? Why do we have to leave home? Why can't Saddam fight like a man? If he did, we knew we could beat him.

Hungry and tired, I slowly closed my eyes until I fell asleep.

I woke up the next morning. It was early, but I could see people already packing to leave. Some had left at sunrise. People did not know their fate. They wanted to get out of the Kurdistan Region of Iraq as fast as they could. People who had been prisoners or faced horrible treatment from the Iraqi soldiers knew there would be no mercy. They would drop chemical bombs on us first. Then they would raid our villages, killing, torturing, and taking prisoners with them. These were brutal soldiers, savages, just like their leader Saddam.

I was in pain inside. I had so much anger. I wanted to stay and fight. There was no way to fight chemical gas though. I thought about my family: my mother, my siblings, Beri, and my newborn son. They needed me. Who would take care of them if I had stayed? Besides, one person fighting won't bring any change. I had no choice but to follow the others and take my family to safety. We had lost the fight with Saddam. Not because we were not warriors or that we did not fight back. We lost because his weapons were bigger and better; his army was much larger than the Peshmerga, the fourth largest in the world to be exact.

Saddam didn't like to play by the rules. It was hard for us to gain our independence with small arms compared to what Saddam had: jets, helicopters, tanks, big bombs, chemical weapons, and a large army.

The second day was nearing its end. We were worn out. We had arrived in Zire Ze, which was the border crossing into

Kurdistan Region of Turkey. It seemed like there was more of the Turkish army than there were of us. They brought in tanks and heavy weapons on their side of the border. Hundreds of people had managed to cross and camp on their side of the border before they had arrived with a large force. This was a smart thing to do. Saddam could kill us in Iraq but not in Turkey. We had to cross and kept on moving with the others.

The Turkish soldiers were telling people in front of us to stop. They were firing into the air to scare people. Despite the bullets in the air coming from the Turkish soldiers to scare the crowd back into Kurdistan Region of Iraq, people kept on walking through the border. They pushed the Turkish soldiers away. Some of the people who tried crossing the border got beat up, but there was just too many of us for the Turkish soldiers to handle. We knew that we were already dead if we did not cross that border, so people just kept on pushing themselves through.

As soon as the Turkish soldiers saw that people just camped on their side of the border they came to a calm. They had thought our people were all going into Turkey to live there. They were hardheaded like Saddam's army, but today it wasn't our war against them, even though they had issues with their own Kurdish population. The Turks didn't gas their people. We were being gassed and we had nowhere else to go. Besides, we were in the Kurdish region of Turkey, not in Ankara or Istanbul.

We just wanted to be safe. We wanted to cross the border so Saddam's army could not get to us and harm our people. I wanted to bring my family to safety. I slowly pushed others in front of me to walk a little faster. I could see that the Turkish army was bringing in more reinforcements and deploying them along the border and among the refugees. I realized that when they could control all the refugees on their side of the border, they would close the border down.

I started hollering, "They are going to close the border; they are going to close the border. Hurry! Please, please hurry!"

Women started to scream. Children started to cry. People started carrying the elderly and little children to be able to walk faster. I saw bushes growing thick by the trees nearby. I untangled my weapons and threw them into the bushes. The Turkish soldiers would single me out and keep me from crossing if I were so heavily armed. I continued to walk with the people in front of me.

In less than an hour, thousands more of us had crossed over into Kurdistan Region of Turkey, including me and my family. A few thousand still remained on the Kurdistan Region of Iraq border when the Turkish soldiers finally closed the border. They started to head towards the Kurdistan Region of Iran border. They had no other choice.

Many did initially go that route, including some of my family members from Brifka. They had hollered at us to go with them to Iran, but we did not want to take that risk. We had been close to crossing the Turkish border and I wanted safety for my family as quickly as possible. We were not free, but for now we were safe. That is all that mattered: we were safe today.

We had been forced out of our homes

and now we didn't know what our fate would be, neither tomorrow nor in the future.

CHAPTER 11

Losing Berivan

The forty thousand Kurds who had managed to cross the border into the Kurdistan Region of Turkey were split into three refugee camps. Some people were sent to Diyarbakir camp, which was the location of the ancient Kurdish capitol, Amed, prior to World War I. Others were sent to Mush Refugee Camp. Then there was Mardin Refugee Camp, where my family and I were sent.

The Turkish government loaded us onto trucks and buses packed with people. They drove us to these places without rest. The Turkish government was paranoid that we would sneak more deeply into their country illegally and influence their Kurdish minority.

It was about a two-and-a-half-hour drive to Mardin. My family and I were pushed into a truck that was packed with other refugees, crying children, and whining elderly complaining about their aches each time the truck hit a bump. The women were getting impatient and saying, "Why is this taking so long?" Little children had accidents and needed to be changed, but there was no place to do it. It was not a fun trip to Mardin Refugee Camp. It is something that I will never forget.

When we arrived, it was a horrible scene. We were given one tent for two families to share. For most of us there were around twenty people in each family. People complained and asked for more tents, but the Turkish soldiers' response was a shove to go away.

Each day our situation got worse and worse. Berivan said to me one day, "Shivan, let's go to Diyarbakir. My family is there. We can help each other. Besides, they live in apartment buildings which are better than the tents we are in."

I told Berivan, "No. We have only been

here a month. We will wait it out. Things might change here, because there are a lot of people at this camp compared to the other two camps in Diyarbakir and Mush."

Beri and I probably could have changed camps, but what about the rest of my family? Who would take care of my mother and siblings? It was going to be hard for all of us to leave Mardin Refugee Camp together. Besides, the Turkish authorities would not allow it. We would need more than one vehicle to take us there, and only the Turks could provide vehicles. So reliable transportation to take us to Diyarbakir would be impossible. How would my mother and siblings be able to take care of themselves here in Mardin? I could not leave them behind.

We were surrounded by Turkish soldiers called Jandarma. They weren't very nice to us. During our time at Mardin Refugee Camp, each time we went up to the Jandarma, to ask them for permission to leave the camp to get supplies, the answer was always no. We had been at the camp for over two months at this point, and were still not allowed to leave. We were like prisoners of war.

Two months later, several leaders of the camp from various villages in Berwari Bala worked tirelessly with the Turkish government at the camp to allow the men to go out and look for work with a permission slip. It did not go easy, and they were turned down each time. Finally, seeing the situation the camp was in made them change their minds. From then on, a few dozen men from our camp in Mardin were allowed to go work for the day. Some work permits were allowed for one day, some for one week, and the others for several months.

When people were allowed to work, the situation of our camp improved. Some people opened markets in the camp, which brought supplies from outside of the camp that were not part of the food rations provided by the Turkish government. Some people opened cafés, and I was one of those people.

I worked outside the camp for about a year and saved some money. My mother had brought my father's savings with her and gave me that money as well. I also had a pistol that had some value when I was a Peshmerga, which my mother had managed to smuggle into the camp with her. I sold it for a good price.

With this money, I opened a theater/café. There weren't that many TVs in the camp. What I did was bring in a TV that was in color, which was rare, and I brought lots of movies from the Turkish cities. The movies were all in Turkish, but regardless, people had some sort of entertainment.

They enjoyed themselves, playing cards and dominos at my café, drinking tea, and talking politics while smoking their cigarettes, watching various Turkish movies and other foreign movies that were translated into Turkish.

Slowly, our people were building the camp their way and not relying on the Turkish government. We managed to make our own water pipeline into the camp, so that we did not have to walk miles to get water. We opened markets, cafés, and schools for our people. We managed to use our skills to make various personal necessities such as clothing and other items that we sold to the Turkish citizens. Our life was getting better since the first day that we had arrived there.

Two years had passed by. We had managed to bring in more tents for families so living space was not so crowded anymore. There were plenty of supplies and those who were working were doing okay. The condition of the camp was still bad because we were still living in tents. We had no paved roads into the camp, everything was gravel and dirt. There was electricity, but we were limited to so many hours a day. We managed to build new toilets for the camp, using the current septic tanks already in place, but the Turkish government rarely brought in a pump truck to clean it and most of the time it was overfilled with feces. It was hard to manage because they did not install a sewage line for us. We had gained a lot on our own, but if the Turkish government had helped us, we would have been able to live our daily lives in somewhat normal conditions.

The clinic at the camp relied on former Peshmerga doctors, of which there were only two. The others were Nurse Practitioners and did not have the proper training to help the thousands of refugees being seen at the camp. Every day there were hundreds of people in line to see the two doctors. Without the proper medicine, there was little that they could do for most of the people at the camp. The doctors brought this up to the Turkish government several times but nothing was done. The only thing that the Turkish government did was to bring in their own doctors with medicine and supplies, which happened just once a month. When people knew that the Turkish doctors were there with proper medical supplies, thousands lined up to see them. This was not a life you would want to have, nor should any person be treated this way.

We had been forced out of our homes and now we were prisoners of a country that did not want us there. Many attempts were made to send us back to the Kurdistan Region of Iraq. Saddam made a deal with the Turkish government that if they sent all his people back, he would forgive those of us who ran off. A few families fell for this trick because of the bad situation at the camps, though it was only a few hundred families. The thousands more Kurdish refugees stayed right where we were, despite the conditions in the camps. We did not believe Saddam.

Most of us were Peshmerga. We would be imprisoned or executed. There was no way for us to go back until the international community gave us a safe corridor into the Kurdistan Region of Iraq.

The end of the third year at the camp was very tough for us. There was an incident at the camp that made our situation worse. A few young men had snuck out of the camp one day. After arriving at a nearby village, they were confronted by a villager and had an argument with him, who was a little older. The older man had a heart attack after arguing with the young men and died instantly. The young men came running back into the camp. That is when the Jandarma found out that it was a few young men from Mardin Refugee Camp who had caused the older man's death.

The Jandarma tried to come into the camp but people did not want that. After an argument broke out the Jandarma started shooting into the air. People started running back into the camp area.

I was at my tent and after hearing the loud noise, I wanted to see what was going on. As I stepped out of the tent, I was struck by a few bullets. I was bleeding all over and passed out right away.

Next thing I saw was people carrying me to the medics. I was put into a Turkish military ambulance and taken to Diyarbakir. I stayed at the hospital for about two months. The doctor repeatedly told me he needed to cut my arm off. I refused even though he said that my arm would be disabled and I would not be able to use it. I told the doctor that I would take my chances.

I was also struck in the rib area, but luckily, the bullet did not go in deep. After a few months, I was released with my arm hanging to my side and not functioning. It was a sad situation. I did not know what to do after that. How was I able to work and take care of Beri and the rest of my family?

After this incident happened, the Turkish government's rules, restrictions, and supplies got worse and worse for us. For weeks, people did not get their food rations, men were not allowed to work outside of the camp, electricity was cut off for days, and water lines were shut off. The Turkish government had done everything they could think of to drive us out of the camps and go back into the Kurdistan Region of Iraq. People were patient though, and many were praying that I had not seen pray before. Many others that were praying stopped believing in anything and thought it was the end and no one cared about them.

The fourth year of living in the refugee camp came. The American government, along with other European countries, established a No-Fly Zone in the Kurdistan Region of Iraq. This was a good thing for the Kurds, as it would prevent Saddam from bombing the Kurdish region. An international Kurdish border was officially set, but this didn't mean Saddam was gone.

The first Gulf War happened, and all of us Kurds watched patiently, hoping for Saddam to be removed from power so that we could go back home. We waited for those countries that truly believed in democracy to remove a tyrant from power; a tyrant who killed so many people, hundreds of thousands of who became victims of his regime, including my own father; a tyrant who used everything in his power to kill off those who opposed him, Kurd and Arab alike.

The Gulf War was brief. It seemed like the war happened for only certain reasons. Saddam was still in power and his people still lived under a tyrant. The Kurds still had no future, even with the no-fly zone. We stayed at the camp because we knew something would tick Saddam off again. Once this happened, there would be another genocide and we would be running for our lives again. No one wants to leave their home. When one person has a stone and the other has a jet, the fight is not fair.

We had had no choice but to flee, and now we had no choice but to stay away. "Since we are here now, we might as well just stay," we told ourselves. We want a future to look forward to for our families. For my siblings, I had to stay. Now that I had children of my own, I have to do all in my power to make sure that they are safe, despite thousands of refugees that went back after the No-Fly Zone.

It is now September of 1991. Beri is

three months pregnant with our second child. We are expecting our second child to be born in March. We are very excited and looking forward to this great day. Another thing that is exciting is that I have applied for refugee status in the United States of America. I really hope that our family is chosen. We don't have a future in Iraq.

I want to raise my family where they will be safe. My siblings are still young, too. I have made the application for them to resettle in the United States as well. The United States of America is a powerful country and we would like to be part of something great. Hopefully, we will be granted refugee status and can start a new beginning.

Many families in Diyarbakir, Mush, and our camp in Mardin are going back to the Kurdistan Region of Iraq. Turkey openly stated that they cannot host us anymore. Out of the forty thousand refugees from Duhok state, only a couple thousand of us are left in the three camps. Many have been granted refugee status in the United States, Canada, Europe and Australia.

My café is not doing as well as it was

a couple of years ago. I am not making money and I am going to close it in a few days. I am now relying on work outside of the camp to support Beri, our son Welat, my mother, and siblings. Luckily, I was able to save some money but I have spent most of it going back and forth to Diyarbakir to fix our refugee resettlement paperwork.

Beri keeps telling me, "I don't want to go, Shivan. I don't want to go to a different country. Please, let's go back to the Kurdistan Region of Iraq. My family has left already, and we should too."

I keep telling Beri that we have no future in Iraq.

We have received messages from some of our family members who went back to Kurdistan Region of Iraq. They have told us that our villages are ruined, our houses and fields are burned to the ground. There is nothing left for them in the villages and they are starting from scratch. I had begged my uncles and grandparents to apply for refugee resettlement to any developed country, that it didn't have to be the United States of America, but they had refused and gone back, only to find that there was nothing for them to go back to.

In the city of Zakho, life is only a little better for them. With so many sanctions on Iraq, it is affecting the Kurdistan Region as well. There is a shortage of supplies and food, the poverty rate is really high in the cities, and living in the village is probably better for most families who are able to farm and start again, if they can find the tools and supplies to do so.

I told Beri, "Please just wait for a few more months. If we don't get our refugee resettlement and are forced out of the camp, then we will go back."

Berivan agreed, but was not happy. She was pregnant and needed medical treatment. She was sick all the time and we did not have the proper medical care for her. I was afraid that if we went back to the Kurdistan Region of Iraq, our situation would be worse. A few times out of the month, I was able to take her to Diyarbakir. The trip was more than a two-hour drive. The doctors there were very expensive and getting medicine drained our pockets. Going back and forth to Diyarbakir was very costly for us.

I tried to get a work permit from the Turkish authorities at Mardin camp. The work permits were given just for the day, and not many were given. Most days, by the time I got the work permit, it was already afternoon. Driving out of the camp and finding work came close to the end of the day, where I had to come back to the camp or I would be violating my work permit.

Most days, I did not have time to find work. It was a tough situation, but I knew deep in my heart that God would help my family and me. He would help our situation improve soon.

January came and it was a new year,1992. Mardin Camp felt like a ghost town. Out of the thousands of people, there were only a few hundred people left. Most of them, like us, were waiting on their refugee resettlement decision paperwork. My mother was sad because of our situation. We were in the middle of winter, it was cold, and we did not have proper tents to cover us from snow coming into our tents. Luckily, when most people left, we had bought extra blankets and foam beds from them to help keep us warm.

One morning, my mother said, "Shivan, son. Are you sure this is what you want? All of our family has left for the Kurdistan Region of Iraq. Maybe we should follow them. I know it is going to be hard, but we can try to survive just like them."

I said, "Mother, I know that we will have a better life in America. There is nothing to go back to. You already know that our house is burned down and so is our farm. We cannot live in the city. There are too many people and the situation is worse than the camp here. Please, just be patient."

My mother continued, "Shivan, I trust your decision. Look at our situation, though. Your siblings are hungry and cold, your wife is pregnant with another child. We can't afford to go to Diyarbakir for medical care every month."

I said, "You are right mother, but you believe in God, right? You are a very religious person. I see you praying more than five times a day. You are always asking God for help. Maybe you should pray for us to hear some good news soon."

She raised both of her hands into the air with her palms out to the heavens and said, "Inshallah! Inshallah, kore min!" (God willing! God willing, my son)

Beri was seven months pregnant now. It was harder for her to do daily chores. When I was able to and not working, I helped as much as I could, as did my mother. The water at the camp had been shut off since there were just a few hundred families left. I was walking with my mother on a daily basis over a mile just to get water. Our electricity was shut off as well. We relied on a fire pit and lanterns for light.

I tried to get supplies with the money I made from working, but restrictions on the camp got tighter and tighter, day by day. We were relying mostly on the rations provided to us.

The people at the camp were scared of eating the food coming in. Our bread had been poisoned before, and a few people had died from it. That was the rumor in the camp even though the Turkish authorities denied such claims. Later, evidence was found that some bread had in fact been poisoned, by God knows who.

Beri was getting sicker and sicker each day. There was not much I could do. I wanted to help her, I wanted to take her pain away. At night was the worst. She would scream loudly and all I could do was hold her hand and comfort her until morning came. That was when it was the quietest. I barely had any sleep at night, trying to comfort my son, Welat, from crying and Beri from the pain that she was going through.

The middle of January came. One afternoon while in line to get a work permit. I heard someone holler my name. "Shivan! Shivan Salih!" I looked up and by the Turkish authorities' office, at the front of the work permit line, was one of the refugee leaders who saw to the needs of the people at the camp. I raised my hand, while keeping my place in line.

"That is me. I am Shivan. I am Shivan Salih."

He said, "Please come here," in Turkish.

I went to the front of the line and greeted him.

He greeted me back and said, "This letter is for you."

I looked at the letter and on the front

was a writing I did not understand. I knew it was English, because it was in a Latin script, and I knew that it was not in Turkish because I had picked up the Turkish language while at the camp. I turned the letter around as it had Arabic script on it as well. I could read Arabic from school, and I started reading the letter.

It stated that the letter was from the Office of Refugee Resettlement. Berivan and I, along with our son and newborn, were accepted for refugee status and we would be resettled in the United States on March 21, 1992. I started to cry and could not stop.

A few people from the camp who knew me came up to me and asked what was wrong.

I told them that I had been accepted to resettle in the United States. I told the Turkish manager of the camp, "Thank you. This letter is going to change my life, thank you."

I hurried to my tent to give Beri and my mother the good news.

I saw them sitting outside of our tents, which were next to each other. I said,

"We have been granted refugee status. We are getting out of this hell hole."

Beri got up slowly, smiling, and gave me a hug.

My mother said, "Piroz bit, kure min" (congratulations, my son).

I told my mother, "Don't worry. I am not leaving without you guys. If you and my siblings don't get approved, then I will cancel my status too. Don't worry, Mother. You have a good heart and your names will come soon."

We did not have much at our tent but gathered what we had with the food supplies and decided to cook a good meal to celebrate. It was a good day and one that I will not ever forget.

One month later, at the end of February, my mother, Beri, the children, and I were sitting in front of our tent one morning. A man came and asked, "Do you know where I can find Aysha Tawfiq?"

My mother replied, "I am Aysha Tewfiq."

He looked at her, smiled and said, "Congratulations! You are going to America."

My mother got up and asked him, "Are you sure? Are you sure that it is my name?"

I grabbed the paper from the young man and looked at it. Indeed, it was the same letter that was sent to me a month prior, but it had my mother's name along with my siblings names on it. I said, "Yes, Mother, it is your name and my siblings names."

My mother became very happy. Though one thing was that they were going to be resettled in September of 1992, six months after our departure. We celebrated once again with what we had for food supplies in our tents.

The next morning, I woke up extra early. I was going to start working for the next two weeks straight, if I was given a work permit. I would save that money along with what we had and give it to someone to look after my mother and siblings until they reunited with us in the United States.

Every morning the first week and half of March, I went to sign up for a work permit. Each day I was there extra early and each day I had been given a work permit. I managed to save some money along with what we had from selling personal items and the café money I made over the years. I found a man named Khalil from Berwari Bala during Friday prayer. He was from Qumri village, where my mother was from originally, and he had been a friend of my father's. I told him about our situation.

He said, "I was waiting on my refugee resettlement paperwork as well. We have been approved too, to travel September of this year as well, but instead we have decided to go back to the Kurdistan Region of Iraq. My family and I will watch over your mother and siblings, under one condition."

I told him, sure, anything.

He said, "My son, who is very hard-headed, does not want to return with us. He wants to take this opportunity to go to the United States. He does not want to come back to the Kurdistan Region of Iraq. Can you watch over him there? Treat him like family? He will be by himself and will need a family."

"Of course. I will treat him like a brother."

Khalil looked at me, smiled, and offered a handshake. I shook his hand.

He said, "We will stay at the camp until September, when your mother and siblings leave. Then, we will too."

I thanked him and handed some money to him. He refused to take it at first. I begged him to take it and told him, you will need this, trust me. He agreed and the deal was made.

Our prayers had been answered and God was watching over us.

Later that day, my younger brother Kawa came running to me. He said, "Shivan hurry!" I asked him, what's wrong? What happened? He was almost out of breath and said, "it's Beri. She is about to have a baby." I hurried with him to where our tents were.

There, I saw many women gathered by my tent. My mother came up to me and said, "Shivan, it's Beri's time. She is due any minute." I said mother, who is helping her? Did you send out for a doctor? My mother replied back, "we have a few midwives, son, but Beri is losing a lot of blood." I said, I will go get a doctor. She looked at me with nervous eyes. I knew something was not right. I knew that Beri was not doing okay.

I ran, ran as fast as I could, to get to the Turkish authorities to get help. It was only half a mile but it seemed like forever, liked miles and miles. Thoughts came rushing into my head. I was not able to think properly and was sweating all over.

I finally arrived at the Turkish authority's site. I asked them for help in Turkish. At first, they looked at me confused and were hesitant. I finally calmed down and explained to them what was going on. They called in an ambulance.

In fifteen minutes a Turkish army ambulance arrived. They picked me up and we headed straight for my tent where Beri was. We arrived as fast as we could.

The medics got out of the ambulance and went into the tent. I went with them, and there was a beautiful baby in Beri's arms.

Beri looked at me and smiled, "Shivan, you are here."

"Yes," I told her. "I am here, Beri."

Beri did not look like the Beri I knew. She was different. She had lost her beautiful skin color and was almost pale and blue. Her majestic amber eyes faded in color and were dark with pain and sorrow. I did not know what to say. I knew something was wrong.

The medics said, "Please give the baby to someone" in Turkish.

I took our baby girl in my arms while the medics took Beri's blood pressure. Our baby girl was so beautiful, just like her mother.

A minute later the medics said in Turkish, "We need to take her to the hospital, her blood pressure is very low."

My sister, Sara, was old enough to watch my younger siblings and my son, Welat, while my mother and I went with Beri to the hospital, along with our newborn baby. Beri held my hand all the way to the hospital. While I held her hand, I was thinking about all the times we were together.

Beri said in a low quite voice, "Shivan, I want to name our daughter Aveen."

I could not hold my tears in and one slid down my cheek. I said "Sure, Beri. Aveen and Welat are great names for our kids."

My mother started to cry.

Berivan did not look good and she was trying to close her eyes. I said, "Beri, stay awake. Stay awake, Beri."

The medics rushed to Beri's side, pushing me away. They yelled at the driver in Turkish, saying, "Hayde, hayde!" (faster, drive faster!)

Beri stopped breathing.

The medics tried to give her oxygen, but it was no use.

By the time we had arrived at the hospital in Diyarbakir, Beri had passed away.

My mother started crying louder and louder.

The tears that I had been controlling while Beri was still alive came rushing down my face, blurring my vision. I did not say one thing. I just held our daughter tightly in my arms while the doctors took Beri away.

They tried doing everything to save her, but it was too late.

Beri died on Friday, March 13, 1992.

We took her body to a local mosque in Diyarbakir for proper religious burying. My mother and I decided that it would be best to send her body back to the Kurdistan Region of Iraq. That is where her family was. They would be able to do a proper funeral for her, even if they were no longer living in Belizan, even if all of Berwari Bala was rubble . If, God willing, one day Kurdistan became independent, I would be able to come back to pay my respects at Beri's grave, to visit her every day.

We spoke with the Turkish authorities from Diyarbakir and they agreed to help us send her body back to the Kurdistan Region of Iraq, if Mardin Camp allowed us. We got permission from Mardin Camp the same day and headed to the border of the Kurdistan Region of Iraq. Before we left, they had contacted the border, controlled now by the Kurds, to let our family know about the situation.

We arrived at the border around midnight. My mother and I were not allowed to cross because we did not have the correct paperwork to do so. We stayed with Beri's body until her father and our other relatives were allowed to come pick her body up on the Turkish side of the border.

It was not a moment that I want to remember. I was not only sad, but had so much pain inside. We hugged our family members one by one. I said my last goodbye to my love. We were then escorted back into the ambulance that brought us back to Mardin Refugee Camp.

CHAPTER 12

Looking Back

Eight days went by. My kids and I were at the Istanbul International Airport, along with hundreds of other families who were flying to the United States. We had flown there from the city of Diyarbakir. We were all waiting for about two hours after being escorted by a few employees at the Istanbul airport. We did not know where we were going, but they had us follow them when we all came out of the plane from Diyarbakir. They put me, my kids, and a few of the elderly into a cart and transported us halfway across the airport and said in Turkish, "Sit here."

We sat there and did not move. I think

most of us did not want any trouble with the Turkish authority. We sat there in our chairs with my two kids and didn't move until the two hours passed. My name was called at the gate I was sitting at. They said it in English and Turkish. I knew what they were saying in Turkish but not English. I went up to the front desk and said in Turkish, "I am Shivan."

They said, "are you ready to board the plane?"

I looked at the airline employee for a few seconds. I looked behind my shoulder and looked at them again, evet (yes), I said in Turkish. One of them escorted me to the plane. I entered the plane and saw dozens of other Kurdish families who were at the Mardin Refugee Camp with me. I greeted them and they greeted me back. I was given my seat number along with my kids.

I sat there in my seat. I looked at my kids. Welat was looking tired and was going to fall asleep any moment. Aveen was already asleep. I sat Welat in his own seat and held Aveen in my lap. My arm was wounded and injured from being shot at the camp.

An older Kurdish lady came by me and said, "I guess they want me to sit here."

I said, "Keremke" (welcome).

She greeted me and sat down. We had a little conversation and she offered to hold my baby, Aveen. I told her that I did not want to trouble her. She said, "Please it's fine. You look injured. I will help you."

I was happy, smiled, and gave her Aveen to hold.

After a few minutes of looking out the window, staring outside, I found myself looking back over my memories of Kurdistan. I saw myself in my younger days, admiring the Peshmerga, being inspired by them and wanting to join them. I saw myself in 1984, taking my opportunity to officially become a Peshmerga with the Kurdistan Democratic Party, which was the best thing I did. At the tender age of thirteen, in 1983, I held a rifle to free my people. Since my birth year in 1971, I had been around war. This was all I knew.

Why? Why did my family and I have to go through this?

I heard a voice coming from the

intercom in Turkish. "Please fasten your belts, the plane is about to take off."

I asked the lady next to me to give me back Aveen. She was an elderly lady and I did not want to trouble her. I thought she was probably tired of holding Aveen in her in her arms.

She replied, "Don't be silly. Look at this precious angel, she is just sleeping."

I put on my seat belt and gazed out the window while the plane took off. I dozed off once again, my mind taking me back to past memories.

I looked at the sky and thought about how small we humans were. Why did some people think they were never going to die and compared themselves to God? I dozed off once again.

I remembered going into the villages when the sun went down. Most of us Peshmerga would sleep in the mosque if there was one available. Many nights, when fighter jets or helicopters were roaming the skies, we were invited into the villager's homes. Many mosques were bombed because they knew that the Peshmerga were there. This was a constant routine. We barely got any sleep.

Before the sun rose, we had to be out of the village and heading back into the mountains. We would stay in the mountains most of the time, our locations were mostly by streams. We would stay there until nightfall came again to head back into the villages.

Before we would go into the villages, our commander would repeat the rules of going into the villages: How we were supposed to act, what we were supposed to say, we needed to show respect to everyone in the villages, we needed to be patient when villagers got upset with us, we needed to treat women like they were our sisters; in no way where we supposed to have any contact with them in any relationship.

By the time we got into the villages, most of us couldn't stand on our feet. We would pass out right away until we were ready to head back out. Our energy was drained most of the day from being a Peshmerga. The good thing about all this was that we had plenty of food and weapons. The villagers fed us well and we got lots of weapons after destroying the enemy's bases, bunkers, and transportation.

We did not celebrate holidays much.

But when it came to be Newroz, the Kurdish New Year, that was the best time for us. Around Newroz, which happens at the spring equinox, is when we attacked the enemy most. We also attacked frequently on the 16th of August. This was the day the Kurdistan Democratic Party was established in 1946, and we had to live up to its name. The enemy feared us most on these two holidays. These two days were when the Iraqi soldiers suffered the biggest losses from us Peshmerga.

After Saddam Hussein's Anfal Campaign, we became refugees. We became refugees in Turkey. We had been there for about two years without leaving our camp. There had been about 19,000 people in our camp. There were over eight people in each tent for about one year. After a year of living in one tent with over eight people, we were given another tent. Finally, I had one tent to me and my kids. My mother and siblings shared the first tent we were in.

Thinking about my café at the refugee camp made me smile. How I had been the first one to bring joy to my people and get their mind off the horrible conditions at the campground. I remembered that day when I stepped out of my tent after hearing loud noises from the campground. I remember the bullets going into my body, feeling the sharp burn of being shot in the arm and ribs. My left arm had been shot and my left rib area. I had felt hot, looked at my wounds, and slowly fainted. That was the last thing I remembered until waking up at the hospital.

During my stay at the hospital, the Turkish doctors told me there was no way they could save my arm. "We must amputate it."

I remember telling them in Turkish, "Just kill me afterwards, if you take off my arm." They kept pressuring me, but I refused to listen to them.

I remember when I first applied for a refugee status to come to the United States, I thought to myself *it's never going to happen, but I might as well try my luck*. Most everyone at the refugee camp registered their names. I might as well do it for my family too.

I was granted refugee status after applying, being shot twice, one in my arm and once in my ribs. There wasn't much medicine. The Turkish doctors sent me back into the camp with my wounds. There were no real doctors in the camp, most were Peshmerga nurse practitioners. Before I was shot, I came out of my tent to see what the fuss was all about in the camp. I had only wanted to see why people were screaming and yelling. That is when the bullets had struck me. I was thinking to myself, *how had we survived the camp for four years, from 1988 all the way to 1992?*

I heard the captain on the intercom again. He spoke first in Turkish, announcing, "We will be descending soon," followed by something in English. Probably the same thing. Soon I would have to learn English, but in that moment I continued to gaze off and think about the past.

In the four years I had been a Peshmerga, over a thousand Iraqi soldiers were killed by the Peshmerga around the Duhok and Zakho area. We lost over two hundred of our own Peshmerga, people who were our relatives, whom we knew and broke bread together. *That is what I regret.* I regret those fights. I wish the killings never happened, I wish that both sides had come to an agreement to prevent war. Why did we have to kill each other for nothing?

The Iraqi government should have asked themselves, and us, "How can we make the Kurdish Region better? What do you want, so we can do it for you?" Everything should have been resolved at the negotiating table rather than on the battlefield. Nothing is resolved by the battlefield and bloodshed.

We had no other choice, though. Why did Iraqi soldiers from Basra have to come and tell the people of Duhok what to do? Why did they come with such hate and oppression? How come the Kurdish people of Duhok were not able to go to Baghdad, Basra, or Najaf to tell them what to do? Why?

"Why" is the question that haunts me now.

The proudest moment of my life is still when I became a Peshmerga. I did not waste my life as a farmer, but rather, I did something good for my people. I did them a service. The name of Peshmerga was sacred, and to be part of that was something just great.

In my personal belief, countrymen come before family. Because if you protect your countrymen and they are safe, then for sure your family is safe as well. If you don't have a country, a piece of land, a place of security and freedom, then where can your family be? The family will be without a place, without a home. First one has to defend his homeland, his people, and his lands. Because when the people are free, then the family is too.

CHAPTER 13

Paying it Forward

It is the year 2003. I have joined the United States Military as a linguist during Operation Iraqi Freedom. I am here, stationed with the United States Civil Affairs team in Kuwait, and have been here for over a week. We have been briefed about what is going to happen. The commanders at the base did not tell us when we would liberate Iraq, but to be ready every single day.

On the night of March 20, 2003, rains of bombs dropped into the capital Baghdad and other areas in Iraq. The next day we entered into the country. When we arrived in the capital, Baghdad, the people welcomed us. Surrounding villages threw flowers at us. We had freed them. Saddam Hussein was officially no longer in power. He was in hiding.

I was with the United States Army Civil Affairs Team until August of 2004. I talked to my captain, Powell, at the base in Mosul, where I was primarily stationed. I asked if we could work in the Kurdish region of Iraq. I told him we would be a better asset there and I could help a lot since I was fluent in Kurdish. Captain Powell said, "let me see what we can do. We might have to come to Mosul from time to time, but we will try to focus in the Kurdish region if our request is granted from the higher ranks."

Captain Powell really liked me. I told him, "To be honest with you, I want to help build the Kurdish region in Iraq. I know that this peace won't last long. The Ba'ath party members are going to turn it into 'Islam vs the West' and it's going to turn ugly. I want to be able to see one region prosper. He agreed and said, "I will let you know tomorrow."

The next afternoon, I heard great news from the captain. He said, "We are going to Duhok state, primarily Zakho city bordering Turkey. That is where our safe house will be."

This is how our civil affairs team ended up in the Kurdistan Region of Iraq, even though we had to go to Mosul once every few months. I was still very happy because we were mostly going to be in the Kurdish region. Working in Kurdistan was bringing old memories once again.

Things were going well, and we were in the initial phases of starting a new Iraqi government. Then, after several months had gone by, chaos happened just as I had predicted. During Ramadan of that year, insurgent attacks increased day by day and Ramadan was a good month used by the insurgents to get recruits.

Despite the attacks, the greatest news came to us on December 13, 2003, when we had heard about the capture of Saddam Hussein in a small rat hole by the United States Army's 4th Infantry Division and members of Task Force 121. His own family members and bodyguards gave the United States military intel of his whereabouts.

A few days after Saddam's capture, all hell broke out. Terrorist snipers began killing our troops, , IEDs (Improvised Explosive Devices), were blowing up the coalition forces. Al Ansar al-Sunnah terrorists and other groups, including former Saddam's Ba'ath party members, tried to squash the democracy we were trying to build for the Iraqi people.

They were taking innocent hostages, mostly foreigners, especially journalists. They were taking them and cutting their heads off while all this was recorded on video and distributed in the Iraqi markets and through other media channels. This was a tactic used by them to frighten the people so that they would not work with us in trying to build a new Iraq.

It is December 20th, 2003. We are driving with a convoy from the Kurdish region, Duhok area, to Mosul when we hear on the radio, "All military vehicles be on the lookout. IEDs are everywhere." A person can make it from Duhok down to Mosul in about 30 minutes driving a regular vehicle. Our Humvees were slower. It usually took us about an hour and half. That day it took us more than four hours because of how cautious we were.

We arrived at our base in Mosul later

that day. We heard about all the IEDs that had blown up the United States military Humvees and other military vehicles. I also heard news about my friend Kawa Tirwanishi; we had served together as Peshmerga. We had both volunteered at the same time as linguists during Operation Iraqi Freedom. He was stationed in Baghdad.

One day their Humvee was blown up, along with others on their team. They lost brave American soldiers that day. We heard about Kawa's bravery as insurgents were closing on them, how he was fighting off the insurgents despite being severely injured.

I called Kawa a couple of weeks later while he was in the hospital recovering from his injuries. We talked for a while and I wished him wellness and a speedy recovery. He explained to me what had happened. He said, "We were driving and suddenly, boom, boom, boom. There was a lot of dust everywhere. I knew that I had been injured, but didn't know how bad. I saw soldiers who were severely hurt. Some were missing their limbs, screaming. Some were dead and not moving. I was on the ground, bleeding.

Each breath I took was harder and harder. I saw my captain had also been severely injured. I heard gun shots and I knew it was not friendly fire. I saw insurgents getting closer to us. I grabbed my captain's rifle and started shooting at them. I called on the other surviving soldiers. They came closer to me and we fired at the insurgents, keeping them away until backup arrived."

I told him, "I know, Kawa. The whole military knows about your bravery. My captain Powell asked me if I knew you and I told him of course I did. I told him that we are friends and were Peshmerga together in the early 1980s."

It is May of 2004. Captain Powell calls me in. He says, "Shivan, I am sorry to say this, but my tour of duty has come to an end."

I looked at him, my face creased with sadness. I said, "Captain Powell I understand." Inside, it really hurt me. I did not want Captain Powell to leave. We had built a strong bond, we helped rebuild Iraq, we helped the Kurdish people, we hired locals to protect us. Through me,

Captain Powell had come to love the Kurdish people.

Later that evening, we called all of our team at the safe house, including locals who had been my friends and relatives whom we had hired to protect us. We talked for a long time.

Captain Powell asked me, "Shivan, what is your last message, or what can I do for you before we leave tomorrow?"

I looked at him and said, "Captain Powell, I know my team cannot be replaced. My last request is that you transfer me to Duhok to be with another Civil Affairs Team. Here is my message for the American people, when you go back:

"I wish that people understood each other. All people must accept one another, no matter who you are, your religion, what color you are, or how you think. Everyone must accept each other. That is the message I want you to take back to the United States."

I continued trying to explain. "An example is that we have a big family back in the United States. We are around ten households. One of our close friends is not blood but we treat him as family and call him brother. My brothers and I treat him as a stepbrother. This is an example in this situation. It shouldn't be a factor to turn your back on him and everything that the family does that the stepbrother has to also. Sometimes, the family has to accept what the stepbrother wants too, and how he thinks differently. We must treat the close friend no different than a family member."

Powell and I, along with others from our unit, were hanging out and just talking. I continued to speak for my unit.

"When I arrived in America, it was tough. I had to really work my way into becoming a United States citizen. We must be accepted in the United States, as if we were born there and our families had been here since the birth of the country. I must be given all opportunities. Whatever I am capable of, I must be given a chance to pursue it. The mentality can't be like, oh you are Kurdish so you can't have the same opportunities as White European Americans. They cannot say, oh, just because there are only fifty thousand Kurds in the United States, none of you can have the same privilege as the majority because we are over three hundred million.

"For example, in Iraq, the Arabs are twenty million and the Kurds are seven million. Are we alike? The answer is no. There is much prejudice and discrimination. If a person from a Kurdish city is capable of being the president of Iraq, then let them be, and vice versa if they were living in Baghdad. When they talk about a united Iraq, then all voices must be heard, no matter the religion or ethnicity.

"That is not the case, though. They tell us, you are Kurds and are not part of Iraq. They say, you are Muslim but not really. Then, when it comes to election time, they say Arabs and Kurds are all Iraqis. That Kurds are our Muslim brothers. When politics take over, and nationality does too, they tell us 'You are not Iraqis but rather Kurds, and neither are you Muslim'. I don't know how my people will be accepted in Iraq, Captain. When we say we are Iraqis, they say, 'No you are not.' When we say we are Muslim, they say, 'No you are not.'

"In Turkey for example, if anyone besides Erdogan becomes the leader, they will do a better job. Even a child is capable of doing a better job as a leader than Erdogan. What is the meaning of war? That is all Erdogan wants to do, take his people into war. There are around twenty-five million Kurds in Turkey. Instead of fighting with them, he should say, okay come to the table. What do you want? What is wrong with one part of Turkey becoming a Kurdish-controlled Kurdistan? Three other pieces of Turkey will still be controlled by the Turks.

"Another example: my younger brother loves politics, and if he runs for office of the United States government in the future, what is the big deal that he is being stopped from seeking this opportunity? How is he going to change anything without the people's voice or other elected candidates? What is so bad if one person from one city or state runs for office? How are things going to be so different?

"So, in Turkey, why are the Turks the only ones to hold every office? Why aren't Kurds allowed to run for office or be elected officials in their own towns and cities? Why does it have to be a Turk all the time? The fact of the matter is they don't accept Kurds. In the four countries of Turkey, Iraq, Syria, and Iran, they don't look at Kurds as citizens of their countries, nor as real Muslims.

"When minority religions and ethnic groups aren't accepted in Iraq, what do they do? They pick up arms. When it comes to the Kurdish women, they can take them as sex slaves without it being a sin. They can take the Yezidi and Christian women without it being a sin, in the name of their religion. Look at Iraq about twenty to thirty years ago, and see how many Christians were here. Now look at Iraq and see how many Christians are left? Not even a million. Look at Turkey and see: how many Christians were there? Now you barely see any, because most of them have been kicked out.

"In these four countries, when minority groups and religions aren't accepted, then they pick up arms. For example, if the Turkish government says to the Kurds, okay, your language will be second to Turkish, you can have schools in Kurdish, and you can be free to practice your culture, I guarantee you many things would change instantly. Why does a Turkish person from Ankara or

Istanbul have to be the mayor of the Kurdish city of Diyarbakir? Aren't Kurds intelligent enough to hold office in their own cities?

"The Turkish government blames the Kurdistan Workers Party or PKK. Well, if you gave Kurds their basic rights in Turkey, there would be no PKK fighting with the Turkish government. Same situation with Iraq: the Kurds would work to unify a central government. We are talking about rights. If Kurds are given more freedom and rights in Iran, Iraq, Syria, and Turkey, the conflict between Kurds and the central government of these countries would end.

"How difficult is it to come to a resolution? It is not difficult. The sad reality is that these three races for over a thousand years have lived their lives this way. They have oppressed minority religions and ethnic groups. Look at the Arab governments. They are unified in the name of Islam. They attacked Persia, and turned it into Islam. They attacked Turkey, and turned it into Islam. They attacked Kurdistan, and turned it into Islam. Many people were forced to become

Muslim. It was by the Arab swords and not their peaceful message of Islam.

"I am not saying I am unhappy that I am a Muslim, but we were forced to it, Captain. Then they spread their religion to Africa, Asia, and tried to spread it into Europe, with the sword. The Arabs were around twenty to thirty million. After forcing so many countries to turn to Islam, they expanded their population using these ethnic groups in the name of Islam. If Europe had not stood in the way of Islam, the whole world would be Islamic today. The Arab governments' agenda was simple: they were to occupy every piece of land in the world and turn all their people into Muslims. The question is, why force others to be something they don't want to be?

"After coming to the United States, Captain, I felt alive. I finally knew I was a free person for the first time in my life, that I had a future. Upon arriving in America, I understood what humanity was all about, what being a patriot meant, and understanding the value of life. America taught me that everything had value. In my homeland, I had no value, except among my own Kurdish people.

"I wish that Kurdistan was a state of the United States. I wish that it wasn't neighbors with Arab states. I wish that Kurdistan was neighbors with the United States or Canada.

"The reason I volunteered to be a linguist was to help the Kurdish Region of Iraq during this Operation Iraqi Freedom, Captain. I joined to help an area that had potential to gain freedom from a dictator, Captain. I wanted to serve both of my homelands. I wanted my fellow Americans to know that Kurds are trustworthy people, that they are loyal."

Captain Powell looked at me and said, "Shivan, you are loyal. You are a good person. Keep talking, soldier."

I continued to talk for Captain Powell.

"After I served with the United States military, I could see the joy of the United States soldiers working alongside Kurds. How happy they were, fighting alongside them. When the US soldiers were in Kurdistan, they were in peace. I am a proud United States citizen who supports our President, George W. Bush. I voted for him, Captain. I voted for him. I will always be grateful for him and as long as the Bush family runs for office, I will vote for them.

"George W. Bush did the right thing, Captain. My assignment with the United States military has mostly been civil affairs, fixing the infrastructure, building schools and hospitals, providing the necessary equipment and supplies needed by the people to run their society. I am still happy because I can try to help rebuild. We have done a lot, right, Captain?"

Captain Powell had tears in his eyes and just nodded.

"Yes, we supplied the local fire departments and police force with the equipment that they needed to do their jobs. We went to every known department in the Kurdistan Region of Iraq, asking them what they needed in terms of supplies and security. Whatever they asked of us we gave them, Captain, we provided that to them. They should be happy we did this, Captain."

"Do you think we will see each other again?" I asked him.

Captain Powell looked at me and said,

"Yes, hopefully we can all reunite again in the future."

"I want to work with the United States military for two more years, Captain. I want to save up for my kids' future," I told him. (The pay is really good.)

"When I was a Peshmerga, Captain, I lost many friends. My best friend Ahmed Ormany, who was my neighbor, lost his life as a Peshmerga. Even though he was 15 years older than me, we were good friends. He was martyred on Newroz, March 21st, 1984. We attacked an Iraqi checkpoint. Most of the Iraqi soldiers fled when we fired upon them. There was one soldier, though, who fought back with grenades.

"When Ahmed got close to the checkpoint, the Iraqi soldier threw the grenades at him and that is how he died. In the end, the Iraqi soldier was also killed by our gunfire. I don't regret anything. I am glad that I was a Peshmerga. What I wished for, then and now, is that we were all one people of Iraq.

"If only the government of Iraq had given us Kurds more autonomy and more rights, I would not be in this situation, Captain. They should have given us the chance to see that we were all one, living in a unified country. If they would have said, 'Okay, you are Iraqi' and would have seriously treated us that way, things would be different."

Our conversation came to an end around midnight. Captain Powell came to me and gave me a big hug. I hugged him back. The other members of our unit came up to me and shook my hand, thanking me for everything. Others in our unit did the same. We said our goodbyes and that was the last time I saw my Civil Affairs team in Zakho.

After working with these guys and the close relationship that we had built, it was hard for me to work with another unit. I was transferred to a team in Duhok City. We worked in the Kurdish region for a few weeks and they said we had to transfer to Kerkuk. It was hard working with this new team. It was nothing like the relationship that I had built with Captain Powell's unit.

I chose to resign my position a few months later.

Afterword

The story of Shivan and Berivan is based on a true story of two lovers from the Kurdistan Region of Iraq during the conflict between the Kurdish people and Saddam Hussein's Ba'ath Regime in the 1980s. Shivan and Berivan were childhood friends from Belizan. a village one hour away from Duhok City in Kurdistan Region of Iraq. They knew each other since childhood. They become lovers after Shivan came across Berivan following an accident he had near the village. Shivan started to have feelings for Berivan and their love ignited from there.

Both of their families were farmers. Shivan wanted to become a Peshmerga, a Kurdish freedom fighter, at the time of Saddam Hussein's reign in Iraq. He lost his father at a young age. Before his father was killed during the conflict, he had encouraged Shivan to focus on his studies rather than being a Peshmerga. Shivan insisted, despite being young, on becoming a Peshmerga. He'd rather fight to free his people than focus on his studies.

The story is unique in many ways. Two lovers that have strong feelings for each other but cannot share them because of cultural norms. They keep their love secret until Shivan asks for Beri's hand in marriage.

The Peshmerga are brave Kurdish fighters, and always have been. They are worth talking about. People need to know their story. With so little resources to protect their people, the Peshmerga manage to create a big impact on the Ba'ath regime and the entire geopolitical region; they have made impressive strides toward the changes they want to see for their people. Like the story of David and Goliath in the Holy Bible, the Peshmerga were David going against the Iraqi army, which was Goliath, being at that time the fourth largest army in the world. The Peshmerga are freedom fighters who fight to the death.

We hope that you enjoyed this book and enjoyed learning about the story of Shivan and Berivan, *War Can't Stop Love*. This book is based on a true story and actual interviews of former Peshmerga and civilians who supported the cause. As each interview came to an end, tears invariably streamed down the interviewees' faces.

About the Author

Newzad Brifki is a Kurdish American author. He arrived in the United States as a refugee with his family at the age of seven in 1992, from the Kurdistan Region of Iraq. Newzad and his family fled during Saddam Hussein's chemical gas attacks in the year 1988. The Kurdish genocide, known as "Al Anfal," was when millions of Kurdish people were forced to become refugees and had no choice but to flee the Kurdistan Region of Iraq. Newzad's family, along with a few thousand more Kurdish refugees, arrived at Ellis Island seeking a new beginning in the early 1990s.

This is Newzad's second book. His first book, *My Journey to America, A Kurdish American Story,* talks about his family leaving the Kurdish Region of

Iraq to find a new beginning. His love for his roots and people has inspired him to write his second book, a beautiful story of two lovers during war, which is based on true events during the Kurdish Iraqi conflict in the 1980s. We trust that you enjoyed this book and enjoyed learning about the life of two young lovers during the conflict between the Peshmerga and the Iraqi army.

Long live all the freedom fighters!

مالپەرێ یێ فەرمێ وارگەهێ بەرواریا

Serê Beraşê 10/4/1984
Foto: Rizgar Kêsteyî

Pîreka- Zaxo 1984 Foto: Rizgar Kêsteyî

Baniya Hesneka- Etrûş
1984
Foto: Rizgar Kêsteyî

Foto: Rizgar Kêsteyî

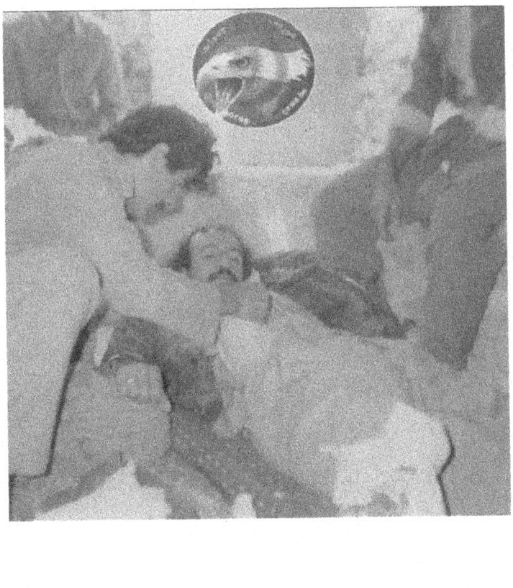

Made in USA - Kendallville, IN
67768_9781737584308
03.15.2023 1326